Progress in Optical Science and Photonics

Volume 35

Indexed by Scopus

The purpose of the series Progress in Optical Science and Photonics is to provide a forum to disseminate the latest research findings in various areas of Optics and its applications. The intended audience are physicists, electrical and electronic engineers, applied mathematicians, biomedical engineers, and advanced graduate students.

Rajib Biswas
Editor

Optical Techniques for Assessing Food Adulterants

Springer

Editor
Rajib Biswas
Department of Physics
Tezpur University
Tezpur, Assam, India

ISSN 2363-5096 ISSN 2363-510X (electronic)
Progress in Optical Science and Photonics
ISBN 978-981-96-8141-9 ISBN 978-981-96-8142-6 (eBook)
https://doi.org/10.1007/978-981-96-8142-6

This Springer imprint is published by the registered company Springer Nature Singapore Pte Ltd.
The registered company address is: 152 Beach Road, #21-01/04 Gateway East, Singapore 189721, Singapore

Preface

Optical techniques have been immensely developed over the years. Their application in diverse applications has also boomed of late. With the advent of development of newer tools and computation platforms, optical techniques have been able to cross over many limitations. One prominent application of optical techniques is in the field of food adulteration. Food adulteration has become a global problem. Proper assessment of these food adulterants and their subsequent remediation remains a vital issue. Toward proper diagnosis of food adulterants, the use of optical techniques has been growing rapidly since last decade. In this context, we endeavor to give a comprehensive glimpse of optical techniques in the realm of food quality and safety with a strong unification of theory and practice.

The book is attempted to be harmoniously balanced, with equal emphasis on theory and practices. We have taken great care to curate each chapter to make it more illustrative and easily intelligible, even if the book covers many areas of plasmonic probes with an intended bridge between theory, instrumentation, and applications. As a result, it is the product of many contacts and a productive conversation process, and it is intended for a broad range of users, including scientists, experts, students, and early-stage researchers.

Optical Techniques for Assessing Food Adulterants provides a forward-looking perspective on the transformative role of optical technologies in safeguarding our food supply. Building upon the foundational knowledge presented in the preceding chapters, we delve into the cutting-edge advancements that are poised to revolutionize food safety surveillance.

The preceding chapters have laid a strong foundation by comprehensively exploring the multifaceted challenges of food adulteration. The first chapter, "Introduction to Food Adulteration," provides a crucial overview of the various forms of food adulteration, their implications for public health, and the urgent need for robust detection methods. The second chapter, "Pesticides, Foodborne Pathogens, and Their Impact: A Comprehensive Study," highlights the significant public health risks associated with the presence of pesticides and foodborne pathogens in our food supply, emphasizing the critical need for effective detection and mitigation strategies.

Chapters "Spectroscopic Techniques for Detecting Food Adulterants"–"Photonic Crystal Sensor for Fat Measurement in Adulterated Milk" delve into the specific applications of various optical techniques in detecting food adulterants. The third chapter, "Spectroscopic Techniques for Detecting Food Adulterants," explores the power of infrared spectroscopy in identifying and quantifying adulterants based on their unique molecular vibrations. The fourth chapter, "Advancements in Fiber-Optic-Based SPR Biosensors for Detecting Food Adulteration," showcases the potential of surface plasmon resonance (SPR) biosensors for highly sensitive and specific detection of foodborne pathogens and other contaminants.

The fifth chapter, "Optical Techniques/Probes for the Detection of Antibiotic Residues in Milk and Dairy Products," highlights the importance of detecting antibiotic residues in the food chain. The sixth chapter, "Recent Advances in Nanomaterial-Based Optical Probes for the Detection of Food Adulterants," explores the exciting possibilities of nanotechnology, particularly the use of nanomaterials as optical probes for sensitive and selective detection of various contaminants. The seventh chapter, "Bridging Technology and Food Safety: An Overview of Smartphone-Based Fluorescence Sensing for Allergen Detection," demonstrates the potential of integrating advanced technology with everyday devices, such as smartphones, to enhance food safety surveillance.

The eighth chapter, "Detection of Adulterants in Soy Based Products Using Non-destructive Spectral and Spectral Imaging Techniques," explores the application of spectral and imaging techniques for detecting adulterants in plant-based foods, highlighting the importance of non-destructive analysis in preserving food quality. The ninth chapter, "A Study on the Detection of Toxicity and Adulteration in Food Products and Proposing a Simple Detection System Using Metal Oxide Semiconductor (MOS) Based Electronic Nose," delves into the application of electronic noses, a novel approach to food safety assessment, utilizing metal oxide semiconductors for detecting volatile compounds associated with adulteration.

Finally, the tenth chapter, "Unveiling the Synergy of Nanocolorimetric and Plasmonic Based Techniques Towards Sensing of Food Adulterants," explores the synergistic potential of combining different optical techniques, such as nanocolorimetry and plasmonics, to enhance the sensitivity and specificity of food adulteration detection. The eleventh chapter, "Photonic Crystal Sensor for Fat Measurement in Adulterated Milk," showcases the application of photonic crystal sensors for specific applications in food analysis, such as determining fat content in milk. The book ends with the twelfth chapter "The Future of Optical Techniques in Food Adulteration Detection" where the critical role of interdisciplinary research and collaboration among scientists, engineers, technologists, and food safety authorities is emphasized. By fostering a collaborative environment, the development and implementation of these innovative technologies can be beefed up—ensuring their accessibility and affordability for all stakeholders across the global food system.

Editing a contributed volume necessitates extensive collaboration between authors, reviewers, and editors. Apart from a few rejections from authors requested, the experience has been very fulfilling. We would like to sincerely thank all the authors who took the time out of their hectic schedules to write their insightful thoughts. We also owe a debt of gratitude to all the referees who helped us revise the contributions with their careful review and recommendations. We do think that the well-chosen names and their equivalents will be of great use to the readers of this book. We believe that by thorough discussion, we have been able to unlock the enormous potential of plasmonic probes.

Tezpur, India Dr. Rajib Biswas

Contents

Introduction to Food Adulteration

Kaushik Kokil Nath, Prakash Nandi, and Pohar Borah

Abstract Since the dawn of human civilization, adulteration in foodstuffs has been a significant concern, as this not only reduces the quality of the food product but also results in a number of side effects on human health. In other words, food adulteration can be defined as the addition of adulterants either intentionally or unintentionally, which raises the quantitative value of raw or prepared food. This adulteration can take place by adding adulterants, removing valuable food components, and substituting it with ingredients with comparatively cheaper substances only for monetary gains and profits. The consumer in this context has suffered various ways during the buying and serving of adulterated food because the food lacks on the part of valuable nutrients; unsafe; may cause health issues along with monetary loss for the consumer. Testing of food requires authority for the evaluation and detection of adulterants in various food products to protect consumers against the deceitful activities of some manufacturers. The present book chapter will attempt to differentiate a number of adulterations happening in various food products and correlated health risks due to the association of this adulteration, along with the different detection methods available at present.

Keywords Food adulteration · Food products · Human health · Nutrients · Food components

K. K. Nath (✉)
Laboratory for Plasma Processing of Materials, Department of Physics, Tezpur University, Tezpur, Assam, India
e-mail: nathkaushik72@gmail.com

P. Nandi (✉) · P. Borah (✉)
Applied Optics and Photonics Lab, Department of Physics, Tezpur University, Tezpur, India
e-mail: prakashnandidpu@gmail.com

P. Borah
e-mail: poharborah50@gmail.com

R. Biswas (ed.), *Optical Techniques for Assessing Food Adulterants*, Progress in Optical Science and Photonics 35, https://doi.org/10.1007/978-981-96-8142-6_1

1 Introduction

Food is regarded as a primary source of energy, and every living organism, whether big or small, requires energy for their survival and to maintain life and growth. Food contains essential nutrients, including carbohydrates, fats, proteins, vitamins, or minerals. Given the importance of food and the demands of food products, food adulteration and fraud have become common practices [1, 2]. Food adulteration refers to the intentional or unintentional contamination or alteration of food products, which can be harmful or of lower quality, compromising the safety, authenticity, and nutritional value of the food [3]. Food fraud involves the deliberate addition of fake substances, removal of genuine substances, or substitution with cheaper, potentially harmful substances for economic gain without the knowledge of the consumer [4]. Mislabelling and misinformation regarding a certain food item also fall under the crime of food fraud [5]. This dishonesty can manifest itself in various such as substituting one ingredient for another (e.g., selling farmed fish as wild-caught), diluting products (e.g., watering down milk), and using counterfit packaging [6]. False claims about a product's provenance, quality, or safety are also included [7]. It also includes false information about the product's origin, manufacturing, or expiration date [8]. The practice of food adulteration and food fraud dates back to the ancient Roman Empire, where laws addressed wine and olive oil adulteration [9]. In the current globalized food supply chain, advancements in manufacturing and transportation have made detecting food fraud increasingly challenging [10]. Food fraud is a significant issue affecting the global food industry and its stakeholders, leading to economic losses, breaking of consumers' trust, and various health issues including kidney diseases, tumors, cancer, diabetes, etc., including fatalities among children [11]. The main goal of food fraud is financial gain. Major factors that contributing to food fraud include globalization, complex supply chains, high demand for specific food items, and inadequate regulation and enforcement laws [12].

These days, it is rather challenging to identify a food market segment that is free from adulterants. Therefore, it is very crucial for us to recognize common adulterants and their hazardous effects on health, such as formaldehyde in fish (Agriculture Times, 2018) [13] and undeclared allergens added to food products (FAO and WHO, 2017) [14]. For combating the threats related to food, it requires a collaborative effort from governments, the food industry, and consumers. With improvement in rules and regulations and their implementation, technological innovation, industry collaboration, and increased consumer awareness, it is possible to reduce the risks and ensure the integrity of the global food supply chain. As the addition of adulterants is increasing day by day, their detection and traceability in food items have become critical. Advanced technology and analytical tools have advanced to the point that the detection of adulterants and fraudulent products is possible, which will be discussed later in this chapter. A few daily essential commodities that are susceptible to adulteration are milk, cooking oil, packaged food items, meat, eggs, fish, fruits, spices, etc.

This chapter, provides an overview of the history of food fraud, types of adulterants, commonly food items that are frequently contaminated, their impact on the economy and health, prevention strategies, and a few detection techniques.

2 Understanding the Concept of Food Fraud

Food fraud dates back to ancient Rome and Athens where there were laws regarding the adulteration of wines with Flavors and colours [15]. Although the practice of food fraud has been prevalent since ancient times the level of fraud and the impact is unmatched compared to present scenarios.

2.1 Defining Food Fraud

Although the definitions of food fraud vary from person to person, the majority of definitions agree that food fraud is an intentional deception for economic gain using food. Within the definitions for food fraud, the motivation or driving factor for economic gain is generally agreed upon. Food fraud definitions from some academic literatures are mentioned below:

According to Spink and Moyer (2011), "Food fraud is a collective term used to encompass the deliberate substitution, addition, tampering, or misrepresentation of food, food ingredients, or food packaging: false or misleading statements made about a product for economic gain" [16].

According to Bouzembrak et al. (2018), "Food fraud in the food supply chain can arise as a result of misrepresentation associated with product integrity (e.g., counterfeit products, expiration dates), process integrity (e.g., diversion of products outside of intended markets), people integrity (e.g., characterizations such as cyber criminals and hacktivists), and data integrity (e.g., improper, expired, fraudulent, or missing common entry documents or health certificates) of information accompanying the food item throughout the supply chain" [17].

According to FAO (2021), "Food fraud is commonly described as any suspected intentional action committed when a food business operator intentionally decides to deceive customers about the quality and/or content of the food they are purchasing in order to gain an undue advantage, usually economic, for themselves" [14].

2.2 Types of Food Fraud

Now that we understand what food fraud is, the key concern is how it is carried out. As the rules and regulations became stringent, people are adopting innovative techniques to commit food fraud. According to the GFSI (Global Food Safety Initiative), there are seven different kinds of fraud, as the following Fig. 1 illustrates [18].

Dilution: The method of dilution is often used by food fraudsters to increase product volume by using a less costly component, thereby boosting profit margins. This may jeopardise the food's originality, safety, and quality. As for example, adding water to milk [19–21] can be thought of such practices.

Substitution: Another popular trick used by food fraudsters is the substitution of high-quality, expensive products with low-quality, cheap products. This may deceive customers, put their health in danger, and create financial losses. This includes adding urea to milk, melamine to milk, red brick powder to spices, etc. [20, 21].

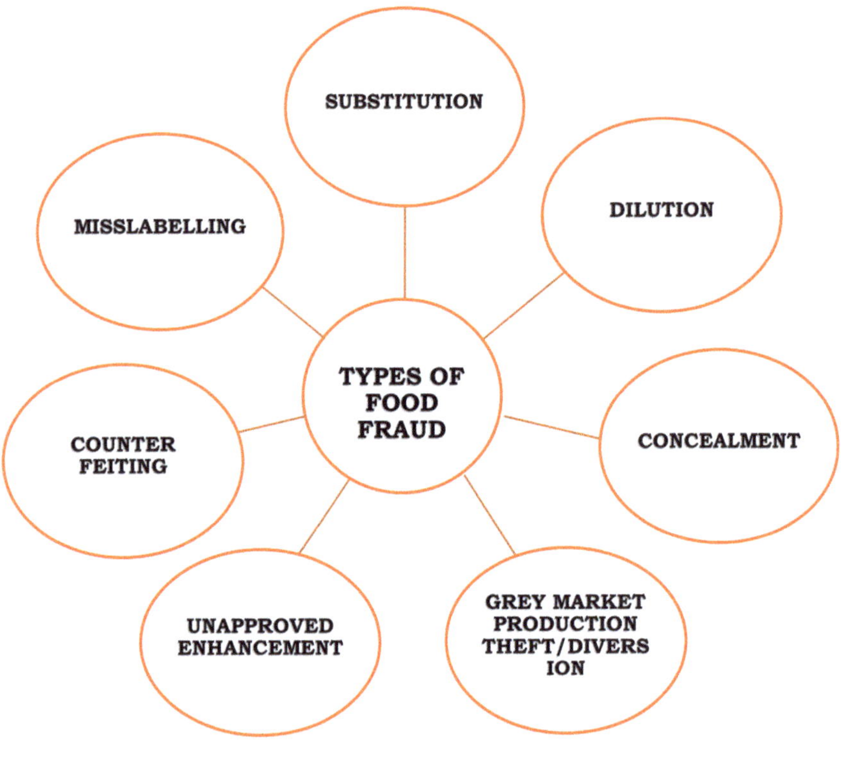

Fig. 1 Diagram illustrating different types of food fraud

Concealment: Concealment involves hiding inferior, substandard, or harmful ingredients in food products. This is a very dangerous practice that could pose a serious threat to customer's health [19, 21].

Mislabelling: In cases of food fraud, mislabelling refers to the practice of giving false or misleading information on food labels to customers regarding the nature, safety, origin, or quality of the product [21–23].

Unapproved enhancement: Unapproved enhancement refers to the use of unapproved chemicals, additives, or processing techniques to enhance a product's flavour, texture, appearance, or shelf life. This poses a serious threat to customers' health [19, 20].

Counterfeiting: Counterfeiting in food fraud involves the creation or sale of duplicate products that are presented as genuine, often copying the brand name, packaging concept, recipe, and processing method of the reputed product [19, 20, 24].

Grey market production, theft, and diversion in food fraud involve the unauthorised production, distribution, and sale of food products outside of authorised channels, regions, or areas. This may harm the reputation of the brand or company [19–21].

Apart from that some common food items and their adulterants and health issues are given in Table 1.

3 Historical Context and Prevalence of Food Adulteration

3.1 Historical Context of Food Adulteration

The adulteration in food, where there is an act to degrade the quality of food with an addition or substitution of lesser quality substance, is of prime concern in history. Several historical records show that the practices have been an ongoing process since ancient civilizations. For example, in ancient Rome, it is documented that Pliny the Elder Natural History from 77 to 79 AD provided several methods for the adulteration of wine, bread, and other staple foods [25]. This gives a very clear indication whereby adulteration in food has been recognized note-worthy at public health and economic issues.

During the mediaeval period, food adulteration went widespread in Europe due to the absence of the regulatory framework. Spices, were commonly adulterated with cheaper substances since they were one of the most expensive commodities [26]. During the Renaissance period, as the expansion of trade reached its boom, instances related to food adulteration increased [9]. Additionally, boom in urbanisation and industrialization leads to a sudden increase in demand for food, creating new opportunities for corrupt practices [27]. In the early twentieth century of the United States, in the famous case of "Poison Squad," Dr. Harvey Wiley experimented with

Table 1 Different food items and some commonly used adulterants and their impact on human health

Selected foods	Adulterants	Impacts on public health
Milk, Water	Starch, urea, and extraction of fat, melamine	Digestive system disorder
Sugar	Chalk powder	Stomach infections
Tea	Black/Bengal gram dal husk with color, artificial pigments/dye and iron fillings	Liver disorders and cancer
Coffee powder	Tamarind and date seed powder and sawdust	Diarrhoea
Salt	White powder, stone, and raw	Stomach disorder
Chili powder	Artificial colors, brick powder, and Sudan dye	Blood and lung cancer
Turmeric	Lead chromate, sawdust, and metal yellow	Carcinogenic
Mustard seed	Seeds of prickly poppy	Argemone Epidemic dropsy and glaucoma
Black pepper	Dried papaya seed	Cardiac arrest and injurious to health
Food grains, pulses, etc	Sand, stone, marble chips, and filth, Lathyrus sativus	Damage digestive tract
Butter	Margarine and starch	Food poisoning
Honey	Fructose syrup/cane sugar	Stomach disorder
Sweets juices	Coal tar dye/metal yellow	Cancer
Green chilies, green peas, and other vegetables	Malachite green	Cancer
Vegetable oil	Argemone mineral oil	Heart disease, skin infection, and cancer
Ghee	Ghee essence, vanaspati, sweet potato, mashed potato, and starch	Cancer and acute renal failure
Carbonic drinks	Aluminium leaves	Asthma and lung disorder
Rice and wheat	Mud grits, soapstone bits, sand, and ergot, melamine	Cancer and genetic mutations harm the human reproductive system
Seafood	Mercury and arsenic	Stomach and brain disorder
Jaggery	Washing soda and chalk powder	Vomiting and diarrhea

Source [4–6]

the use of preservatives in food with volunteers to draw people's attention to extensive use of adulterants in food and their frightful association with human health [28].

3.2 Early Regulatory Efforts

The earliest significant regulations with respect to food adulteration emerged in the nineteenth century. The Adulteration of Food and Drink Act was passed in the United Kingdom way back in the year 1860, which can be said to be one of the earliest legislative efforts made in this direction to combat food adulteration [29]. It was succeeded by the Sale of Food and Drugs Act-1875, due to which tightening regulations ensued [30]. The Pure Food and Drug Act of 1906 in the United States was a direct result of public outcry and truly laid the groundwork for contemporary food safety standards [31].

3.3 Prevalence of Food Adulteration in Modern Times

With several efforts to combat adulteration in food dating back several hundred years, it remains one of the most widespread issues globally. In developing countries with weak regulatory frameworks, high economic pressures coupled with inadequate law enforcement result in a widespread of food adulteration. For instance, some disquieting levels of adulterants in dairy products, spices, and edible oils have been revealed in studies conducted in India. According to a study published in Food Control in 2013, it was found that 70% of milk samples were contaminated by various products like benzoic acid, urea, water, skimmed milk powder, sugar, detergent etc., in India [32]. Moving towards the developed countries, even though the laws and regulations are more practiced, adulteration of foods is still present and often driven with monetary profit.

One of the notable instances of food adulteration in this modern era is the Horsemeat Scandal in Europe, 2013, where the public is being cheated by selling them horsemeat in the products labelled as beef [33]. Adulteration of honey with sugar syrups has been a thorn in the European Union and United States markets, where according to a report by the European Commission's 14% of the honey samples that were tested had adulterants [34].

3.4 Socio-Economic Impact

Adulteration in food has profound implications for the socio-economic sector, which works not only affect the health of the public but also raises concerns with respect to economic stability, consumer trust, and the integrity of practices in the market. It also contains the hidden costs in the form of medical expenses that come along with foodborne illnesses, and includes the added cost of productivity and healthcare needs in the long term [18]. Financially, businesses face losses due to recalls of adulterated products, an answerability to the legal system, and damage to brand reputation; for a small-scale producers experiencing severe financial hardship, a reduction in access to the market, and potential threat to business closure [35, 36]. Adulteration in food reduces the trust of the consumer; disrupting the supply and demand chain which later impacts the dynamics of the market [37]. Disruption of international trade results in an import ban and economical losses as shown by melamine scandal in China, and it compromising the regulatory adherence [38, 39]. Socially, the adulteration of food disproportionately impact low-income groups and marginalized sectors of populations, continuing poverty and disparities in health treatment. This places a significant burden to public health systems because of the increase in foodborne illness, while the government is trying to reduce the cost substantially with the enforcement of food safety and regulations [40]. Advanced technology for detection, strict rules and regulations, enhanced awareness about food adulteration among the masses, and emphasis on collaborative efforts by the Government, Industry, and Consumers are the UCs that have to be ensured [41].

4 Impact on Consumer Health

Exposure to food Adulteration is a serious threat to humankind. Consumption of adulterated food containing toxic substances can lead to several diseases, from mild to life-threatening. Widely seen examples of such diseases are diarrhoea, skin diseases, liver problems, cancer, asthma, kidney failure, urinary tract infections, glaucoma, cancer, etc. [18]. In some cases, the manufacturers intentionally add some less harmful and cheaper substances to increase the profit margin. The addition of corn flour to onion powder or other spices, roasted corn to coffee, and vegetable oils to extra-virgin olive oil are some of such examples. These harmless or less harmful substances can be attributed to some allergic reactions in some cases [42]. In this section, we shall discuss some of the most commonly used adulterants, their impact on consumer health, and some of their detection methods.

4.1 Melamine

Melamine is a widely used adulterant in milk to increase the protein content. It is a nitrogen-rich compound. Consumption of melamine can lead to serious health conditions such as renal failure, urinary tract infections, renal calculi, etc. In China, in 2008, newborn milk products were adulterated with melamine, which affected 2,94,000 children, 50,000 hospitalisations, and 6 deaths. After this outbreak, melamine adulteration has become a matter of great concern [43]. In a recent study, melamine was detected by the colorimetric change of surface-modified gold nanoparticles with polythymidine oligonucleotides. This polythymidine oligonucleotide combines with melamine to form a double-strand DNA-like structure, which leads to the aggregation of the nanoparticles. When SYBR Green I (SG I) was embedded into the system, the interaction between polythymidine oligonucleotide and melamine increased, allowing the change of the colour of the nanoparticles from red to dark purple with a limit of detection of 6 $\mu g\ L^{-1}$ [37]. Surface-modified silver nanoparticles were synthesised using polyaniline and employed for the detection of melamine in milk, which showed up to 0.1 μM melamine concentration in daily life milk samples [38]. The formation of hydrogen bonding between melamine and gallic acid, which was used as a reducing and stabilising agent, showed an extremely low limit of detection of 0.012 ppm [39].

4.2 Urea

Urea is one of the primarily used adulterants in milk. Consumption of urea generally leads to digestive system disorders. However, long-term effects include blood disorders, peptic ulcers, liver cirrhosis and failure, heart diseases, bone marrow abnormalities, renal failure, etc. [18, 43]. Urea is commonly detected with enzymatic sensors [40]. However, with the progress of the research, the enzymatic sensors have been replaced by non-enzymatic techniques. One of the examples of urea detection is electrochemical detection using urine-immobilised graphene nanoplatelets and graphitized nanodiamond nanocomposites [41, 44]. Electrospun nanofibers modified with ZnO nanoparticles using a simple electrostatic process can also be used for the detection of urea [45].

4.3 Formaldehyde

Formaldehyde is known as a carcinogenic adulterant. There are several reports of the illegal addition of formaldehyde to food items such as meat, seafood, fish, vegetables, milk, and fruits to extend their shelf life [42]. Elevated consumption of formaldehyde can cause vomiting and diarrhoea. It may also attack the optical nerve. In the long term, it can develop respiratory, digestive, cardiac, nephrological, and neurological problems, as well as cancer [46]. Formaldehyde can be detected by conventional methods such as high-performance liquid chromatography, gas chromatography, mass spectrometry-based methods, etc. [42]. In a recent study, it was detected in corn using a screen-printed electrode-based electrochemical biosensor with the help of cyclic voltammetry. In the same work, formaldehyde has also been detected using a hematite nanostructure-modified indium tin oxide-coated glass electrode with the help of the optical technique [47]. This adulterant was also detected using the fluorescence resonance energy transfer (FRET) strategy and a two-photon (TP) radiometric fluorescent probe [48]. Another very rapid and low-cost method was developed using a heated PDMS (polydimethylsiloxane) microfluidic chip [49]. A low limit of detection of 100 ppb was achieved by using poly-dopamine-functionalized hollow mesoporous silica spheres [50]. To detect formaldehyde in fish, an optical sensor was used using an optical fibre bundle [51]. Several colorimetric methods were employed to detect this adulterant in different food samples using biodegradable hybrid polymer films [52], thin films of tapioca starch [53], thiol-functionalized silica nanoparticles [54], etc. In recent works, different machine-learning algorithms were used to detect formalin in different fruit and vegetable samples [55].

4.4 Lead Chromate

Several kinds of research are going on for lead chromate, a carcinogenic toxic substance used to increase the colour vibrancy of turmeric. This highly toxic substance may cause anaemia, brain damage, and cancer [20].

Erasmus et al. investigated lead chromate with the help of Fourier transform-Raman spectroscopy using both univariate and multivariate statistics and found the limit of detection to be 0.5% [56]. Another study developed a powder X-ray method for the detection of this adulterant in turmeric powder and achieved a limit of detection of 5% [57]. A different article reported portable X-ray fluorescence analysis that could detect lead with a detection limit of 2 mg Pb/kg for spice powders. The other method they developed is a colorimetric assay using diphenyl carbazone [58]. Tiwari et al. used a laser-induced breakdown spectroscopic technique for the detection of the adulterant in turmeric powder [59].

4.5 Brick Powder

Brick powder is a commonly used adulterant for the adulteration of red chilli powder. It can be very harmful to human health. If taken at an elevated dose, this can cause blood and lung cancer [42]. In a very recent study, Sarkar et al. reported adulteration detection of red chilli powder with red brick powder using machine learning techniques [60].

5 Legal and Regulatory Frameworks Related to Food Quality and Safety

Ensuring food quality and safety is a critical global, which leads to the implementation of legal and regulatory frameworks. These frameworks aim to protect consumers by enforcing fair trade practices and upholding the standard for public health. This part of the chapter mainly sheds light on the development of these legal and regulatory frameworks globally for ensuring the quality of food and safety regarding it, with a particular focus on India.

5.1 Global Legal and Regulatory Frameworks

The legal and regulatory frameworks for food safety and quality are diversified and complex across the world. Major international organisations play crucial roles in establishing standards, agreements, and guidelines to ensure food safety.

5.1.1 Codex Alimentarius Commission (CAC)

The Food and Agriculture Organization (FAO) and the World Health Organization (WHO) jointly established the Codex Alimentarius Commission (CAC), with the objective of protecting the health of consumers and ensuring fair practices during food trade. The work of CAC is to develop international standards for food, guidelines, and codes of practice to safeguard of consumer health and to ensure ethical practices in the food trade [61].

5.1.2 World Trade Organization (WTO)

The Technical Barriers to Trade (TBT) Agreement and the Sanitary and Phytosanitary (SPS) Agreement of the World Trade Organisation (WTO) offer the basis for ensuring that food safety standards are grounded in scientific knowledge. The primary goal of this international organisation is to guarantee that trade occurs as easily, reliably, and freely as possible. It deals with global trade regulations [62].

5.1.3 European Food Safety Authority (EFSA)

The primary responsibility of EFSA is to evaluate and communicate risks within the food supply chain to inform and protect consumers. Most of EFSA's work is done in response to particular requests for scientific assistance since its recommendations help risk managers in developing policies and decisions [63].

5.1.4 United States Food and Drug Administration (FDA)

The FDA is responsible for regulating food safety in the United States under laws such as the Federal Food, Drug, and Cosmetic Act and the Food Safety Modernization Act (FSMA), which emphasizes on preventing contamination of food and rapid response to outbreaks of foodborne illness. The Food and Drug Administration is responsible for protecting the public's health by making sure that pharmaceuticals for humans and animals, biological products, and medical devices are safe, effective, and secure. It also makes sure that food supplies, cosmetics, and radiation-emitting products are safe in the USA [64].

These organisations and agreements reflects the united effort globally for maintaining the high standards for food safety and quality, with an aim to make the international trade smoother, and protect consumers worldwide.

5.2 Legal and Regulatory Frameworks in India

With a population of 1.5 billion with a diverse food culture, the country factually faces unique challenges according to the quality of food and safety of human health. In a country like India, the legal and regulatory framework is ruled by three sectors: the laws, regulations, and authorities. The Food Safety and Standards Authority of India (FSSAI) operates under the Food Safety and Standards Act, of 2006.

5.2.1 Food Safety and Standards Authority of India (FSSAI)

The FSSAI has been established under the Food Safety and Standards Act, 2006, which is a unique body concerned and responsible for protection and promotion of public health by regularly monitoring and enforcing the safety of food. FSSAI consolidates various acts and orders earlier handled by different ministries and departments for different food-related issues [65].

5.2.2 Food Safety and Standards Act, 2006

The purpose of this Act is to consolidate the laws relating to food, to establish the Food Safety and Standards Authority of India, so as to regulate the manufacture, storage, sale, and import of food articles in order to ensure availability of safe and wholesome food to human beings, and for matters connected therewith or incidental thereto [66].

5.2.3 FSSAI Regulations

Several rules and regulations have been framed under the Act of FSS, and some of them will be mentioned below to name a few:

- Food Safety and Standards (Food Products Standards and Food Additives) Regulations, 2011
- Food Safety and Standards (Packaging and Labelling) Regulations, 2011
- Food Safety and Standards (Contaminants, Toxins and Residues) Regulations, 2011
- Food Safety and Standards (Prohibition and Restrictions on Sales) Regulations, 2011
- Food Safety and Standards (Licensing and Registration of Food Businesses) Regulations, 2011 [67].

5.2.4 Legal Metrology Act, 2009

It is an Act to establish and enforce standards of weights and measures, regulate trade and commerce in weights, measures and other goods which are sold or distributed by weight, measure or number and for matters connected therewith or incidental thereto [68].

5.2.5 Prevention of Food Adulteration Act, 1954 (PFA)

Though it has now been superseded by the FSS Act, the PFA laid the ground for ensuring pure and wholesome food to the consumers and protect them from fraudulent and deceptive trade in India, significantly influencing the current regulatory framework [69].

5.2.6 Agricultural and Processed Food Products Export Development Authority (APEDA)

APEDA was established during 1985 under the act of Agricultural and Processed Food Products Export Development Authority. The main objective of APEDA is to promote export of the agricultural and processed food items including at least fruits, vegetables and their products, meat and meat products, and poultry and poultry products that meet the international standards [70].

6 Challenges and Opportunities

Despite India having a very extensive legal framework, ensuring food quality and safety remains challenging due to multiple systemic issues. The challenge remains to have very effective enforcement and implementation of laws on food safety due to resource constraint and stretchiness of the Indian food industry, which is huge and diverse.

Implementation and Enforcement: Effective implementation and enforcement of food safety laws remain a significant challenge due to limited resource, inadequate infrastructure, and the vast, diverse nature of the Indian food industry. The scale and complexity of the sector, encompassing small-scale vendors to large manufacturers, complicate consistent regulatory oversight and compliance.

Awareness and Education: Food safety standards and practices are poorly understood by food business operators and consumers, necessitating widespread awareness and education campaigns. Promoting knowledge about safe food handling, adulteration risks, and regulatory requirements is critical to improving compliance and consumer safety.

Technological Advancements: Advanced technologies for monitoring, traceability, and testing can significantly enhance food safety by detecting adulterants and ensuring product integrity. Innovations such as rapid testing kits offer solutions address adulteration effectively.

Agency Coordination: Effective coordination among different regulatory bodies and stakeholders is essential for cohesive food safety management.

7 Conclusion

Food adulteration has deep historical roots, dating back to ancient times, but the issue persist globally in modern times. Food tampering occurred from ancient Rome through the Middle Ages in Europe. It persists today in insidious ways because the world is more interwoven, food travel involves complex supply lines, and consumers demand for food products. This has a significant effect on society and the economy. It makes consumers lose trust in their food, putting their health in jeopardy and disrupting the economy. Economic impacts include healthcare costs for treating illness caused by adulterated food, business losses from product recalls or by removing food product from the shelves, fines, and reputational damage. On the health front itself, adulterated food consumption can cause various sicknesses-ness. These can range from minor issues like rashes to serious conditions such as cancer, kidney problems, or heart disease. Common adulterants added to food are melamine, urea, formaldehyde, lead chromate, and brick dust. The chapter discusses the detection techniques for these adulterants and emphasizes how modern technology makes a difference in ensuring the safety of food. The laws and regulations drawn out at the global and country levels provide a critical framework for effective deterrence of food tampering. Global organizations such as the Codex Alimentarius Commission and national organizations such as the FSSAI prescribe standards, inspect food, and monitor enforcement. These rules can work effectively only if we all contribute to them and have strict and constant checks. As the food fraud increases, enhancement of the detection technology, strengthening regulatory enforcement, and educating consumers about contaminants food become imperative. Collaborative efforts are needed to secure the food supply chain, promote health, and restore consumer confidence in food safety.

References

1. Y. Chen, M. Michalak, L.B. Agellon, Importance of nutrients and nutrient metabolism on human health. Yale J. Biol. Med. **91**, 95–103 (2018)
2. M. Friedman, Nutritional value of proteins from different food sources. A Rev. J. Agric. Food Chem. **44**, 6–29 (1996). https://doi.org/10.1021/jf9400167
3. A. Choudhary, N. Gupta, F. Hameed, S. Choton, An overview of food adulteration: concept, sources, impact, challenges and detection. Int. J. Chem. Stud. **8**, 2564–2573 (2020). https://doi.org/10.22271/chemi.2020.v8.i1am.8655
4. Economic Impacts of Food Fraud | Agricultural Economics. Available online: https://agecon.unl.edu/cornhusker-economics/2020/economic-impacts-food-fraud
5. M.J. Moreira, J. García-Díez, J.M.M.M. de Almeida, C. Saraiva, Consumer knowledge about food labeling and fraud. Foods **10**, 1095 (2021). https://doi.org/10.3390/foods10051095
6. T.F. Leo, *Handbook of Food Analysis*, 3d edn-Volume Set (CRC Press, Boca Raton, FL, USA, 2015). ISBN 9781482297843
7. Z.-Y. Zhou, *Global Food Security* (Routledge, New York, NY, USA, 2019). ISBN 9781315406947
8. J.-L. Xu, D.-W. Sun, *Chapter 8—Imaging Spectroscopic Technique: Hyperspectral Imaging*. Available online: https://www.sciencedirect.com/science/article/abs/pii/B9780128142646000086
9. P.B. Hutt, P.B.I. Hutt, A history of government regulation of adulteration and misbranding of food. Food Drug Cosmet. Law J. **39**, 2 (1984)
10. S.M. van Ruth, P.A. Luning, I.C.J. Silvis, Y. Yang, W. Huisman, Differences in fraud vulnerability in various food supply chains and their tiers. Food Control **84**, 375–381 (2018). https://doi.org/10.1016/j.foodcont.2017.08.020
11. R.S. Hellberg, K. Everstine, S.A. Sklare, *Food Fraud: A Global Threat with Public Health and Economic Consequences* (Academic Press, San Diego, CA, USA, 2020)
12. S.M. van Ruth, W. Huisman, P.A. Luning, Food fraud vulnerability and its key factors. Trends Food Sci. Technol. **67**, 70–75 (2017). https://doi.org/10.1016/j.tifs.2017.06.017
13. B.K.K.K. Jinadasa, C. Elliott, G.D.T.M. Jayasinghe, A review of the presence of formaldehyde in fish and seafood. Food Control **136**, 108882 (2022). https://doi.org/10.1016/j.foodcont.2022.108882
14. FAO, *Food Fraud—Intention, Detection and Management* (FAO, Rome, Italy, 2021)
15. K. Robson, M. Dean, S. Haughey, C. Elliott, A comprehensive review of food fraud terminologies and food fraud mitigation guides. Food Control **120**, 107516 (2021). https://doi.org/10.1016/j.foodcont.2020.107516

16. J. Spink, D.C. Moyer, C. Speier-Pero, Introducing the food fraud initial screening model (FFIS). Food Control **69**, 306–314 (2016). https://doi.org/10.1016/j.foodcont.2016.03.016
17. Y. Bouzembrak, B. Steen, R. Neslo, J. Linge, V. Mojtahed, H.J.P. Marvin, Development of food fraud media monitoring system based on text mining. Food Control **93**, 283–296 (2018). https://doi.org/10.1016/j.foodcont.2018.06.003
18. S. Bansal, A. Singh, M. Mangal, A.K. Mangal, S. Kumar, Food adulteration: sources, health risks, and detection methods. Crit. Rev. Food Sci. Nutr. **57**, 1174–1189 (2017). https://doi.org/10.1080/10408398.2014.967834
19. H.W. Woldemariam, B.D. Abera, The extent of adulteration of selected foods at Bahir Dar, Ethiopia. Int. J. Interdiscip. Res. **1** (2014). ISSN 2348-6775 (online)
20. M. Momtaz, S.Y. Bubli, M.S. Khan, Mechanisms and health aspects of food adulteration: a comprehensive review. Foods **12**, 199 (2023). https://doi.org/10.3390/foods12010199
21. C. Cordella, I. Moussa, A.-C. Martel, N. Sbirrazzuoli, L. Lizzani-Cuvelier, Detection and quantification of honey adulteration by chemometric analysis of some minor compounds. J. Agric. Food Chem. **50**, 1751–1764 (2002). https://doi.org/10.1021/jf011096z
22. S. Charlebois, A. Schwab, R. Henn, C.W. Huck, Food fraud: an exploratory study for measuring consumer perception towards mislabeled food products and influence on self-authentication intentions. Trends Food Sci. Technol. **50**, 211–218 (2016). https://doi.org/10.1016/j.tifs.2016.02.003
23. D.C. Moyer, J.W. DeVries, J. Spink, The economics of a food fraud incident—case studies and examples including melamine in wheat gluten. Food Control **71**, 358–364 (2017). https://doi.org/10.1016/j.foodcont.2016.07.015
24. K. Everstine, J. Spink, S. Kennedy, Economically motivated adulteration (EMA) of food: common characteristics of EMA incidents. J. Food Prot. **12**, 399 (2023). https://doi.org/10.4315/0362-028X.JFP-12-399
25. Pliny the Elder, The natural history, BOOK I.1, Dedication. 1 Lemaire Informs Us, in His Title-Page, *That the Two First Books of the Natural History Are* ed. by M. Alexandre, in His Edition. Available online: https://www.perseus.tufts.edu/hopper/text?doc=Plin.+Nat.+toc
26. E.J.T. Collins, Food adulteration and food safety in Britain in the 19th and early 20th centuries. Food Policy **18**, 95–109 (1993). https://doi.org/10.1016/0306-9192(93)90018-7
27. R. Tannahill, *Food in History* (Paw Prints, 2008). ISBN 9781439506974
28. D. Blum, *Poison Squad: One Chemist's Single-Minded Crusade for Food Safety at the Turn of the Twentieth Century* (Penguin Books, S.L., USA, 2019). ISBN 9780143111122
29. R.E. Ferner, J.K. Aronson, Medicines legislation and regulation in the United Kingdom 1500–2020. Br. J. Clin. Pharmacol. **89** (2022). https://doi.org/10.1111/bcp.15497
30. Sale of Food and Drugs Act 1875. Available online: https://www.legislation.gov.uk/ukpga/Vict/38-39/63/enacted#:~:text=1875%20CHAPTER%2063%2038%20and
31. FDA, *Part I: The 1906 Food and Drugs Act and Its Enforcement* (FDA 2019, 1)
32. R. Chugh, G. Kaur, A study on milk adulteration and methods of detection of various chemical adulterants qualitatively. IOP Conf. Ser. Mater. Sci. Eng. **1225**, 012046 (2022). https://doi.org/10.1088/1757-899X/1225/1/012046
33. M. Kulas, *Policy Responses to Reduce the Opportunity for Horsemeat Adulteration Fraud: The Case of the European Union.* Available online: https://krex.k-state.edu/items/92511994-5ee3-4897-932e-bdd76379b58e
34. V. Kružík, A. Grégrová, A. Rajchl, H. Čížková, Study on honey quality evaluation and detection of adulteration by analysis of volatile compounds. J. Apic. Sci. **61**, 17–27 (2017). https://doi.org/10.1515/jas-2017-0002
35. M.B. Rahman, I. Jahan, T. Parveen, F. Ali, S. Akhtar, M.B. Hossain, M. Mahmud, M.R. Karim, An update on formaldehyde adulteration in food: sources, detection, mechanisms, and risk assessment. Food Chem. 136761 (2023). https://doi.org/10.1016/j.foodchem.2023.136761
36. A. Haji, et al., Selected food items adulteration, their impacts on public health, and detection methods: a review. Food Sci. Nutr. **11**(12), 7534–7545. https://doi.org/10.1002/fsn3.3732
37. W. Cao, S. Yu, X. Chen, J. Liu, X. Wang, Y. Ma, Q. He, Q. He, Z. Tang, M. Wei, Novel rapid detection of melamine based on the synergistic aggregation of gold nanoparticles. Food Chem. **428**, 136789 (2023). https://doi.org/10.1016/j.foodchem.2023.136789

38. M. Waqas, L. Campbell, Polyaniline-coated surface-modified Ag/PANI nanostructures for antibacterial and colorimetric melamine sensing in milk samples. ACS Omega **8**, 24010–24015 (2023). https://doi.org/10.1021/acsomega.3c02627
39. X. Liu, H. Liu, L. Wang, Q. Chen, Q. Wu, In situ functionalization of silver nanoparticles by Gallic acid as a colorimetric sensor for simple sensitive determination of melamine in milk. ACS Omega **6**, 23630–23635 (2021). https://doi.org/10.1021/acsomega.1c03927
40. A.A. Ibrahim, A.E. Eltayeb, E.Y. Salih, H.Y. Kordkheili, N.F. Khair, K.A. Halim, M.J. Haron, M.M. Yusoff, Two-dimensional ytterbium oxide nanodisks based biosensor for selective detection of urea. Biosens. Bioelectron. **98**, 254–260 (2017). https://doi.org/10.1016/j.bios.2017.06.015
41. V. Kumar, N. Sinha, A. Nambiraj, A. Suvitha, H.K. Lee, S. Thakur, A.K. Srivastava, Amperometric sensing of urea using edge activated graphene nanoplatelets. RSC Adv. **5**, 13278–13284 (2015). https://doi.org/10.1039/C4RA12594K
42. M.B. Rahman, I. Jahan, T. Parveen, F. Ali, S. Akhtar, M.B. Hossain, M. Mahmud, M.R. Karim, An update on formaldehyde adulteration in food: sources, detection, mechanisms, and risk assessment. Food Chem. 136761 (2023). https://doi.org/10.1016/j.foodchem.2023.136761
43. A. Haji, K. Desalegn, H. Hassen, Selected food items adulteration, their impacts on public health, and detection methods: a review. Food Sci. Nutr. **11**, 7534–7545 (2023). https://doi.org/10.1002/fsn3.3732
44. V. Kumar, A. Mukherjee, M. Aslam, A.K. Srivastava, Graphene nanoplatelet/graphitized nanodiamond-based nanocomposite for mediator-free electrochemical sensing of urea. Food Chem. **303**, 125375 (2020). https://doi.org/10.1016/j.foodchem.2019.125375
45. F.L. Migliorini, D.P. Santos, O. Fatibello-Filho, Urea impedimetric biosensing using electrospun nanofibers modified with zinc oxide nanoparticles. Appl. Surf. Sci. **443**, 18–23 (2018). https://doi.org/10.1016/j.apsusc.2018.02.168
46. M.A. Al Mamun, S.M. Rahman, K. Alam, M.S. Islam, M.S. Rahman, An overview of food adulterants and their health impacts. Int. J. Sci. Res. Publ. **11**, 780–796 (2021). https://doi.org/10.29322/IJSRP.11.05.2021.p11381
47. M. Kundu, S. Hazra, S. Biswas, M. Mukhopadhyay, B. Mondal, D. Ghosh, Comparative studies of screen-printed electrode based electrochemical biosensor with the optical biosensor for formaldehyde detection in corn. Food Bioprocess Technol. **14**, 726–738 (2021). https://doi.org/10.1007/s11947-021-02604-3
48. G. Yuan, H. Luo, H. Qin, Z. Li, M. Yang, Q. Wang, A novel fluorescent probe for ratiometric detection of formaldehyde in real food samples, living tissues and Zebrafish. Food Chem. **331**, 127221 (2020). https://doi.org/10.1016/j.foodchem.2020.127221
49. X. Weng, Y. Kang, J. Guo, Y. Xing, Q. Sun, S. Chen, Rapid detection of formaldehyde concentration in food on a polydimethylsiloxane (PDMS) microfluidic chip. Food Chem. **114**, 1079–1082 (2009). https://doi.org/10.1016/j.foodchem.2008.10.027
50. J. Zong, S. Wang, Y. Chen, X. Zhang, Q. Ma, D. Han, Z. Ni, Rapid and highly selective detection of formaldehyde in food using quartz crystal microbalance sensors based on biomimetic polydopamine functionalized hollow mesoporous silica spheres. Sens. Actuators B Chem. **271**, 311–320 (2018). https://doi.org/10.1016/j.snb.2018.05.089
51. M. Yasin, N. Yahya, S.W. Harun, H. Ahmad, S.S. Damanhuri, D. Puryanti, M.A. Mahdi, S. Shaari, M.H. Md Khir, Z. Yusoff, Fiber bundle sensor for detection of formaldehyde concentration in fish. Opt. Fiber Technol. **52**, 101984 (2019). https://doi.org/10.1016/j.yofte.2019.101984
52. K. Rovina, N.A. Aznan, S. Siddiquee, Development of biodegradable hybrid polymer film for detection of formaldehyde in seafood products. Sens. Bio-Sens. Res. **27**, 100310 (2020). https://doi.org/10.1016/j.sbsr.2019.100310
53. W. Wongniramaikul, W. Limsakul, A. Choodum, A biodegradable colorimetric film for rapid low-cost field determination of formaldehyde contamination by digital image colorimetry. Food Chem. **249**, 154–161 (2018). https://doi.org/10.1016/j.foodchem.2018.01.021
54. S. El Sayed, A.M. Mahfouz, R. Kamel, A. Khattab, M.I. Khalil, M.W. Helmy, N.M. Zeid, Chromogenic detection of aqueous formaldehyde using functionalized silica nanoparticles. ACS Appl. Mater. Interfaces **8**, 14318–14322 (2016). https://doi.org/10.1021/acsami.6b03224

55. S.P.S. Brighty, G.S. Harini, N. Vishal, Detection of adulteration in fruits using machine learning, in *Proceedings of the IEEE WiSPNET 2021, 6th International Conference on Wireless Communications, Signal Processing and Networking (WiSPNET)* (IEEE, 2021), pp. 116–119. https://doi.org/10.1109/WiSPNET51692.2021.9419402
56. S.W. Erasmus, M. Muller, L. van den Berg, M. Manley, Real or fake yellow in the vibrant colour craze: rapid detection of lead chromate in turmeric. Food Control **121**, 107714 (2021). https://doi.org/10.1016/j.foodcont.2020.107714
57. R. Paranthaman, J.A. Moses, C. Anandharamakrishnan, Development of a method for qualitative detection of lead chromate adulteration in turmeric powder using X-ray powder diffraction. Food Control **126**, 107992 (2021). https://doi.org/10.1016/j.foodcont.2021.107992
58. A.M. Lopez, D.R. Stoll, S.L. Simonich, K.A. Anderson, Assessing analytical methods for the rapid detection of lead adulteration in Tiwari
59. M. Tiwari, N. Kaur, A.K. Rai, P.K. Rai, Laser-induced breakdown spectroscopy: an approach to detect adulteration in turmeric. Spectrosc. Lett. **46**, 155–159 (2013). https://doi.org/10.1080/00387010.2012.702707
60. T. Sarkar, S. Ghosh, S. Pramanik, M. Mukhopadhyay, Artificial intelligence aided adulteration detection and quantification for red chilli powder. Food Anal. Methods **16**, 721–748 (2023). https://doi.org/10.1007/s12161-023-02445-0
61. FAO, CODEXALIMENTARIUS FAO-WHO. Available online: https://www.fao.org/fao-who-codexalimentarius/en/
62. World Trade Organization, What Is the WTO? Available online: https://www.wto.org/english/thewto_e/whatis_e/whatis_e.htm
63. EFSA, European Food Safety Authority – EFSA | European Union. Available online: https://european-union.europa.eu/institutions-law-budget/institutions-and-bodies/search-all-eu-institutions-and-bodies/european-food-safety-authority-efsa_en#:~:text=The%20European%20Food%20Safety%20Authority
64. USAGov, Food and Drug Administration | USAGov. Available online: https://www.usa.gov/agencies/food-and-drugadministration#:~:text=The%20Food%20and%20Drug%20Administration
65. FSSAI, FSSAI. Available online: https://fssai.gov.in/cms/about-fssai.php
66. FSSAI, Food Safety and Standards Act 2006. Available online: https://fssai.gov.in/cms/food-safety-and-standards-act-2006.php
67. FSSAI, FSSAI Regulations. Available online: https://www.fssai.gov.in/cms/food-safety-and-standards-regulations.php
68. Ministry of Consumer Affairs. Home | Department of Consumer Affairs | Ministry of Consumer Affairs Food and Public Distribution | Government of India. Available online: https://consumeraffairs.nic.in/acts-and-rules/legal-metrology/the-legal-metrology-act-2009
69. Indian Railways, CHAPTERX Prevention of Food Adulteration Section A the Prevention of Food Adulteration Act. Available online: https://indianrailways.gov.in/railwayboard/uploads/codesmanual/MMVol-II/Medical_II_Ch10_data.htm#:~:text=Enacted%20by%20the%20Parliament%20in
70. APEDA, Agricultural and Processed Food Products Export Development Authority. Available online: https://apeda.gov.in/

Pesticides, Foodborne Pathogens, and Their Impact: A Comprehensive Study

Pohar Borah, Kaushik Kokil Nath, and Prakash Nandi

Abstract The demand for high productivity and yield in agriculture incorporates the usage of pesticides. However, uncontrolled usage and chronic exposure to these agrochemicals can impact human health, aquatic species, and biodiversity. Foodborne pathogens, on the other hand, can occur at various stages of food production and delivery and can lead to severe outbreaks. This chapter gives insight into pesticides and their application, foodborne pathogens, the impact of pesticides on human health, mostly used detection methods, followed by reported foodborne outbreaks, and lastly usage of optical probes for the detection of the pathogens.

Keywords Foodborne · Pathogens · Pesticides · Optical probes · Foodborne outbreaks

1 Introduction

1.1 What is Pesticide and Its Classification

In agriculture, crops are highly vulnerable to being attacked by different kinds of pests. Pesticides, in this scenario, have become a boon to the agricultural industry and the farmers for higher production and less loss of crops to pests [1]. The increasing population of the world demands more food production and higher yields. Along with several other factors, pesticides are also contributing to higher production of

P. Borah (✉) · P. Nandi
Applied Optics and Photonics Research Laboratory, Department of Physics, Tezpur University, Tezpur, Assam, India
e-mail: poharborah50@gmail.com

P. Nandi
e-mail: prakashnandidpu@gmail.com

K. K. Nath
Laboratory for Plasma Processing of Materials, Tezpur University, Tezpur, Assam 784028, India
e-mail: nathkaushik72@gmail.com

R. Biswas (ed.), *Optical Techniques for Assessing Food Adulterants*, Progress in Optical Science and Photonics 35, https://doi.org/10.1007/978-981-96-8142-6_2

food and reducing the loss of crops to pests [2]. Pesticides can be defined as some compounds or mixtures of compounds that can be used to control or kill different pests attacking the crops. These pests include fungi, rodents, insects, weeds, etc. [3]. Pesticides are also used to control some vector-borne diseases like dengue or malaria [4].

Pesticides can be classified depending on various parameters. Based on the hazards they create; The World Health Organization has classified pesticides into five categories as shown in Table 1 [5]. Based on different chemical structures, different pesticides can be classified as organophosphorus, organochlorines, carbamates, and pyrethroids [6]. Organochlorine pesticides work as hormone receptors and are resistant to environmental degradation. These pesticides such as endosulfan or indane work as effective insecticides [7]. Organophosphorus works as acetylcholinesterase (AChE) inhibitors that lead to permanent overlay of neurotransmitters across a synapse. Parathion and malathion are some examples of this category. Carbamate pesticides are similar to organophosphorus in chemical structure and they work on the principle of attacking the nerve signal which ultimately leads to poisoning and death of the pest. Carbofuran, aldicarb, and methomyl are some examples of the carbamate category. Natural pyrethrins are duplicated for the manufacturing of pyrethroid pesticides. Cypermethrin, deltamethrin, and Permethrin fall into this category and they behave as neurotoxic insecticides [7–9]. Pesticides can also be classified based on the target pest. For example, the pesticides that attack insects are known as insecticides, ones that attack herbs are known as herbicides, fungi attackers are known as fungicides, etc. as shown in Fig. 1 [4, 9]. In addition to this, depending on the work mechanisms, pesticides can also be categorized. Some pesticides attract the pests whereas others repel them. The ones who attract are known as attractant pesticides and the other ones are known as repellents. Some pesticides control the growth of the pests. They are known as growth regulators [8].

Table 1 Classification of pesticides by WHO [5]

Class		LD_{50} for rat (mg/kg)	
		Oral	Dermal
Ia	Extremely hazardous	<5	<50
Ib	Highly hazardous	5–50	50–200
II	Moderately hazardous	50–2000	200–2000
III	Slightly hazardous	Over 2000	Over 2000
U	Unlikely to present acute hazard	5000 or higher	

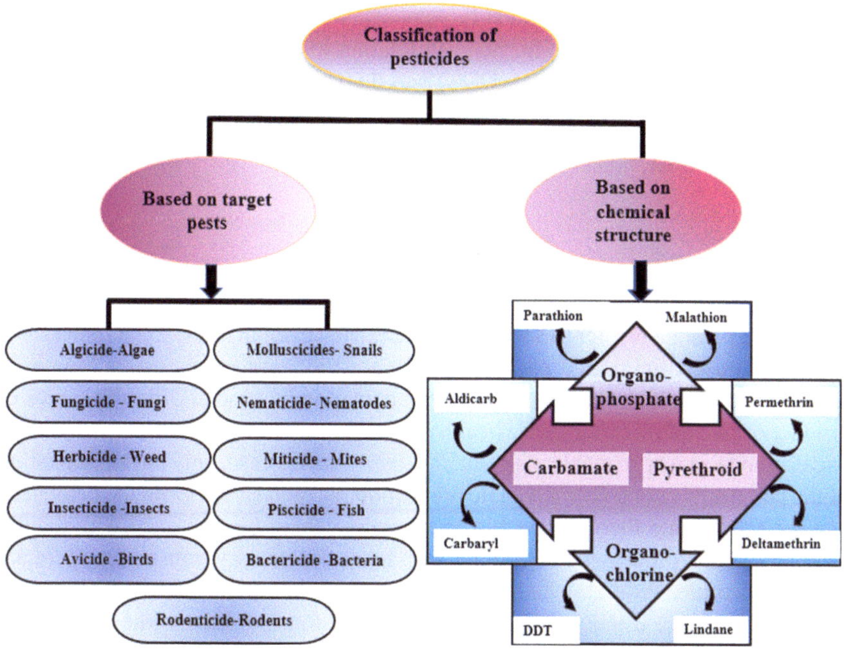

Fig. 1 Classification of pesticides based on target pests and chemical structure [4]

1.2 Applications of Pesticides and Their Historical Background

Despite being beneficial for the agriculture and food industry, uncontrolled application of pesticides has created a serious threat to humankind, aquatic species, and biodiversity [9]. To mitigate these pitfalls, several practices have been brought to light. Different countries have decided the maximum residue limits of pesticides that can remain in food commodities after it is served to the consumers. Another practice that has been applied is Good Agricultural Practice or GAP [10]. Pesticides can be applied in different ways like spraying or dusting etc. Various kinds of nozzles like hydraulic, centrifugal, heat energy, and air blast are used for spraying purposes. Water, air, oil, and soap solutions are commonly used carriers of pesticides. In the case of dusters, dust formulations are made in such a way that the active substance is mixed with a lightweight substance. Dusters also have different categories like getz applicators, power dusters, hand crank dusters, electric motor-powered dusters, etc. [11]. Nowadays artificial intelligence is playing a crucial role in pesticide application. With the advent of technology, automation or drones are being explicitly used in agricultural fields for the application of pesticides [12].

Pesticide usage is not very new. The early usage of pesticides dates back to the period when people burned sulfur to control pests in ancient Rome. Along with that, highly toxic arsenic and mercury were also used as pesticides. The revolution of synthetic pesticides began in 1939. In relation to that, the insecticidal effect of Dichlorodiphenyltrichloroethane (DDT) was discovered by Paul Miller then. From then till now, several pesticides have been synthesized using different compositions and applied in agricultural fields [8, 13].

1.3 Foodborne Pathogens

Foodborne pathogens refer to a broad spectrum. However, in brief, some species like bacteria, viruses, fungi, and parasites that cause foodborne diseases are known as foodborne pathogens. These pathogens can contaminate food at different stages of production and delivery and cause health risks to consumers. In addition to this, some bacteria and fungi are capable of producing toxins. In this case, only the identification of the pathogen is not a very adequate prevention for food safety. Both the pathogens and their toxins are thermally stable and thereby cannot be destroyed with regular food production processes [14]. Along with food, the primary transmitter, there are some other routes through which the pathogens and their toxins can be transmitted. Humans, the host, usually suffer from gastrointestinal problems like diarrhea or vomiting; however, the invasive ones produce other symptoms as the toxins can affect any system of the body [15].

2 Impact of Pesticides on Human Health

2.1 Neurological Disorders

Numerous Investigations have been carried out in recent times to study the impact of pesticides on Neurological illness. According to a study conducted on the agricultural population in Cache County, it was found that exposure to pesticides increases the risk of Alzheimer's and dementia [16]. Parkinson's disease also can be caused by pesticides and their metabolites, which alter xenobiotic metabolism and interfere with mitochondrial function [17].

Organophosphate and carbamate pesticides can lead to overstimulating cholinergic receptors in the nervous system which causes neurotoxic disorders [18]. Some pesticides such as indane and endosulfan alter the function of GABA in the central nervous system which can result in various neurological diseases [19].

2.2 Effect on Reproduction

Different studies have shown that pesticide exposure can lead to erectile dysfunction. Oliva et al., reported that women who worked with herbicides for two years before conceiving were more likely to experience infertility [20].

Neural tube defects (NTDs) are the most frequent and deadly congenital illnesses that affect newborns. A study conducted in China showed that there is a high chance that organochlorine pesticides are one of the major causes of NTDs [21]. It has also been observed that pesticide exposure to parents also increases the risk of birth defects [22].

2.3 Respiratory Disorders

The application of pyrethroids and organophosphates in agricultural fields elevates the risk of the development of asthma in young children living near the area [23]. Experiments have also shown a high risk of lung cancer due to exposure to agricultural pesticides [24].

2.4 Diabetes

Studies have shown that certain organochlorines and organophosphates can interfere with the normal functioning of the endocrine system which can lead to the risk of diabetes. These insecticides mainly target insulin signaling and glucose metabolism [25].

2.5 Cancer

Cancer is a global issue that has been drawing the attention of researchers for a long. Direct contact with pesticides is the main reason for cancer [26].

Several researchers reported that pesticides induce carcinogenic effects by disrupting estrogen receptors or damaging DNA in breast tissues and boosting cancer-causing cells and DNA mutation in females which lead to breast cancer [27].

Certain insecticides and herbicides have genotoxic qualities that can favor cancerous cell growth. Employees who work in agriculture fields especially those handling or applying pesticides are prone to liver cancer [28].

Though the exact mechanism is still under research, Greenop et al., found that infantile brain tumors are caused due to exposure to pesticides throughout the pregnancy [29].

In a very recent work, it has been reported that the risk of having bladder cancer is directly related to the amount of exposure and time spent in the pesticidal environment [30].

Researchers have also found a connection between leukemia and pesticides. According to some research groups, parental pesticide exposure is the main cause of acute non-lymphoblastic leukemia [31, 32].

3 Detection Methods for Pesticides

In modern times, pesticides have become an integral part of agriculture for controlling pests to enhance crop yields. However the residues of these pesticides remain in food products, which pose a significant health risk to humans, necessitating robust detection methods. In this section, we shall provide an overview of the regulations for the use of pesticides on different food products, the different detection mechanisms, and later the authentication methods.

3.1 Regulation of Different Pesticides for Different Food Matrices

Regulation for pesticides varies across different food categories, The risk factors involved with residues of different pesticides are also different. The regulatory framework is designed in a way to protect the health of the consumer while keeping in mind the productivity in agriculture.

3.1.1 International Regulatory Standards

The international regulatory standards for the use of pesticides in farming are mainly given by the Codex Alimentarius Commission, which sets the maximum residue limits (MRLs) for pesticides, to ensure food safety and smoothness of international trade [33]. Based on various risk evaluations these standards are set that are conducted by the Joint FAO/WHO Meeting on Pesticide Residues (JMPR), which is responsible for the evaluation of scientific data on pesticide residues and their potential threat to human health [34]. The standards set up by Codex are recognized globally and serve as a reference for national regulations to coordinate safety measurements around the world.

3.1.2 Regional and National Regulations

Specific regulations have been tailored for different countries because of their agricultural practices and patterns of diet. In the case of the European Union's Regulation (EC) No. 396/2005, MRLs for pesticides are set in food and feed of plant and animal origin. The regulation incorporates regular monitoring with the submission of reports, which ensures the safety of consumers [35]. The U.S. Environmental Protection Agency (EPA) establishes a tolerance limit for pesticide residues and involves precise testing and imposition mechanisms [36]. In countries like India, the Food Safety and Standards Authority of India (FSSAI) sets the MRLs for the residues of pesticides in food [37] using the Food Safety and Standards Regulation, which checks on contaminants, toxins, and residues [38]. Their main work is to ensure that the residues of pesticide levels in foods are within the limits that are safe and permissible for human consumption without any threat to their health [39].

3.1.3 Category-Specific Considerations

Food placed in different categories, such as fruits, vegetables, grains, animals, and dairy products, comes with their unique challenges for their regulation in the use of pesticides. Several factors are taken into consideration, such as the chemical properties and composition of the pesticides, their pattern of degradation, and the methods of processing the food that impacts setting up the MRLs. For example, vegetables often have lower MRLs compared to grains, as they are highly vulnerable to the absorption of pesticides and direct consumption as compared to grains [40].

3.2 Detection Mechanism

For regulatory compliance and safety of consumers detection of pesticide residues with high accuracy plays a crucial role. Different types of detection mechanisms have been developed that involve both traditional and advanced analytical techniques.

3.2.1 Chromatographic Techniques

Chromatography, especially gas chromatography (GC) and liquid chromatography (LC), incorporated with mass spectroscopy (MS), offers very high sensitivity and specificity for detecting the presence of multiple pesticides in different food categories [41]. These techniques mainly involve the separation of the pesticide residues from the components of food, followed by identification and quantification based on mass-to-charge ratios.

3.2.2 Colorimetric Techniques

This technique is widely used for on-site monitoring of pesticide residues due to its ease of use, low cost, and fast results. The methods are mainly based on the reaction of pesticide residues and a specific reagent that produces a color change that can be quantified with the use of a colorimeter or spectrophotometer. There has been the development of colorimetric biosensors and others, like enzyme-linked immunosorbent assays (ELISA) for detecting trace amounts of organophosphate and carbamates in food [42, 43].

3.2.3 Spectroscopic Methods

It mainly includes two spectroscopy methods, infrared (IR) and ultraviolet–visible (UV–Vis) spectroscopy, which are commonly used for pesticide detection. The benefits that come with these methods are that they are safe and non-harmful, and can be used as a pre-test before more detailed analysis. Nuclear magnetic resonance (NMR) spectroscopy plays an important role in helping to interpret the chemical structures of pesticides along with their metabolites [44].

3.2.4 Immunoassays

Immunoassays mainly use antibodies like enzyme-linked immunosorbent assays (ELISA) for the detection of specific pesticide residues. These tests are low-cost, easy to use, and ideal for on-site testing, but offer lower specificity and sensitivity compared to the techniques of chromatography [45].

3.2.5 Biosensors

These are emerging trends in technology for the detection of pesticides with biological molecules, like the use of enzymes and nucleic acids, for the recognition and quantification of the residues of pesticides. Biosensor offers various benefits like real-time analysis, and rapidity in detection, that open up a new field of testing and continuous monitoring [46].

3.3 Authentication Procedures

The process of authentication ensures the accuracy and reliability of the methods for the detection of pesticides. Three processes are mainly involved here, i.e., standardization, validation, and quality control measures that ensure the integrity of the test results.

3.3.1 Standardization

The process of standardization mainly includes consistent development of the protocols for the preparation, analysis, and evaluation of the sample. The International Organization for Standardization (ISO) provides guidelines regarding the analysis, consistency, and comparability of results across laboratories around the globe for residues of the pesticides found in food [47].

3.3.2 Validation

The validation method is mainly used for assessing the performance features of the detection methods, which include sensitivity, specificity, accuracy, precision, and robustness. It mainly ensures that the methods that are being used are fit for the purpose and that the reliability of the detection of pesticide residues is within the limits that are required [48].

3.3.3 Quality Control

The measurement of quality control involves the use of certified reference materials (CRMs) while carrying out audits that are both internal and external. The carried measures help in the identification and correction of errors, which provides assurance, reliability, and accurate analysis of the residues from pesticides in the food [49].

4 Impact of Foodborne Pathogens and Usage of Optical Probes for Their Detection

4.1 Reported Hazards

In history, there have been several reports of foodborne pathogen outbreaks that severely harmed humankind. From 1999 to 2019, there were several fresh produce outbreaks in Europe and North America. There were 25 bacterial outbreaks with 6,553 cases, 140 hospitalizations, and 57 deaths in Europe, primarily due to *E. coli*. North America reported 177 outbreaks with 16,275 cases, 1,785 hospitalizations, and 139 deaths, with the highest number of cases from *Salmonella*. Viral outbreaks in Europe were 7 with 13,923 cases and 2 deaths, while North America reported 17 with 793 cases. In both countries, *norovirus* had the highest number of cases. Parasite outbreaks totaled 12 with 1,760 cases in Europe and North America reported 39 with 5,220 cases, mainly caused by *Cyclospora* [50]. In 2012, Japan had an *E. coli* outbreak causing 110 infections and 7 deaths. The USA, in 2011 reported *Listeria monocytogenes* outbreak resulted in 146 infections and 30 deaths [51]. From 2015 to 2020, China had 410 bacterial outbreaks with 4,991 cases and 3 deaths along with 20 viral outbreaks [52]. India mostly experienced several bacterial outbreaks from 2009 to 2018 [53].

4.2 Rapid Screening of the Pathogens

Foodborne pathogens are responsible for several diseases such as diarrhea, typhoid hemorrhagic colitis, hemolytic uremic syndrome (HUS), etc. These diseases are severe in nature and life-threatening. Therefore, to prevent such dreadful scenarios and to provide a safe food supply, the rapid detection of these pathogens is indispensable [54, 55]. Optical probes are proven to be one of the best measures for this purpose with high sensitivity and rapid detection time. Optical biosensors can be classified into different categories based on the working principle as colorimetric, fluorescence biosensors, surface plasmon resonance (SPR), and surface-enhanced Raman spectroscopy (SERS) based biosensors [54].

4.2.1 Colorimetric Optical Sensors

Colorimetric sensors having fast detection time and easy operation detect pathogens via a change of color that can be visualized by the naked eye or optical detectors. In a very recent work, *Salmonella* Typhimurium was detected using iron oxide (Fe_3O_4) nanozymes that were coupled with Cu and poly(pyrrole-2-carboxylic acid) yolk-shell nanozymes, and the resultant Fe_3O_4@Cu@PCPy YSNs were surface modified with flower shaped aptamer-complementary strand (Apt-CS) conjugates. The proposed aptasensor could detect *S. Typhimurium* with an LOD of 1 CFU/mL [56]. Xiao et al. detected *Listeria monocytogenes* using magnetic nanoparticles coupled with cefepime mediated with branched polyglutamic acid. A platinum-palladium nanozyme was further introduced to the above platform to design the enzymatic colorimetric sensor and coupled with a smartphone for signal access and analysis which showed a LOD of 3.1×10^1 CFU/mL [57].

4.2.2 Fluorescent Probes

For regular monitoring and rapid detection of pathogens, fluorescent probes are widely used. Zhang et al. synthesized silica-core quantum dot-shell nanocomposites. Polyethyleneimine was used to create an interlayer on the silica core after which the resultant nanocomposite detected *Salmonella Typhimurium* with an LOD of 5×10^2 cells per mL [58]. A fluorescent probe was designed using a quantum dot functionalized with receptor binding protein RBP 55 to detect *Salmonella* in food matrices with a detection limit of 2 CFU/mL [59].

4.2.3 SPR Sensors

SPR sensors are label-free sensors that measure the biomolecular interaction between pathogens and ligands. Bhandari et al. developed an SPR assay for the detection of *Salmonella Typhimurium* using *S.* Typhimurium flagellin-specific monoclonal antibodies [60]. In another SPR sensor, *Listeria monocytogenes* was detected using wheat germ agglutinin as the selective ligand which bio-molecularly interacted with the surface carbohydrate component of *L. monocytogenes* [61].

4.2.4 SERS-Based Sensors

SERS-based sensors used for the detection of foodborne pathogens can be classified into two categories viz. label-free and label-based. *Escherichia coli* O157:H7 was detected using a SERS-based sensor where starch magnetic beads coated with gold nanoparticles and surface functionalized with specific antibodies were used. A bifunctional linker protein was used for the immobilization of antibodies which also served as Raman reporter. The LOD of the sensor was found to be as low as a single cell [62]. *S. aureus*, *E. coli*, and *P. aeruginosa* were extracted using a SERS-based sensor having a sandwich-like structure. Non-interfering SERS probes were used for bacterial labeling and, for extraction purposes, ConA-functionalized magnetic nanoparticles were used. They reported the preparation of the PP3 SERS probe which had a very high Raman activity because of which a mixed probe formed with MP1 And DP2 could detect multiple pathogens simultaneously with a LOD of 1 CFU $\cdot$ mL^{-1} [63].

5 Conclusion

Pesticides have been proven to be a cornerstone for the mordern agriculture and food industry in terms of enhancing crop yields and higher production. However, the residue of pesticides that remain in the food matrices because of uncontrolled usage can lead to several diseases. Consequently, robust detection and regulation of pesticide residues are critical for food safety.

Foodborne pathogens and their toxins also present a significant threat to human health. Several bacterial, viral, and parasite outbreaks have been reported globally which infected a large number of population causing significant number of deaths from different parts of the globe. Therefore, the detection of these pathogens is also very important for a safer food supply system. Among various detection methods, optical probes are found to be one of the best measures that can comply with this need.

Acknowledgements Authors P. Borah, K.K. Nath, and P. Nandi would like to thank Tezpur University for the institutional fellowship.

References

1. D.J. Ecobichon, Pesticide use in developing countries. Toxicology **160**, 27–33 (2001). https://doi.org/10.1016/S0300-483X(00)00452-2
2. J. Popp, K. Pető, J. Nagy, Pesticide productivity and food security. A review. Agron. Sustain. Dev. **33**, 243–255 (2013). https://doi.org/10.1007/s13593-012-0105-x
3. I. Mahmood, S.R. Imadi, K. Shazadi, A. Gul, K.R. Hakeem, Effects of pesticides on environment, in *Plant, Soil and Microbes*, ed. by K.R. Hakeem, M.S. Akhtar, S.N.A. Abdullah (Springer, Cham, Switzerland, 2016), pp. 253–269. https://doi.org/10.1007/978-3-319-27455-3_13
4. P. Borah, R. Biswas, Impactful analytical schemes for assessing pesticides in tea: a comprehensive review. Measurement 113505 (2023). https://doi.org/10.1016/j.measurement.2023.113505
5. World Health Organization, *The WHO Recommended Classification of Pesticides by Hazard and Guidelines to Classification 2019* (World Health Organization, Geneva, Switzerland, 2019). Available online: https://www.who.int/publications/i/item/9789240005662
6. K.-H. Kim, E. Kabir, S.A. Jahan, Exposure to pesticides and the associated human health effects. Sci. Total. Environ. **575**, 525–535 (2017). https://doi.org/10.1016/j.scitotenv.2016.09.009
7. F.P. Carvalho, Pesticides, environment, and food safety. Food Energy Secur. **6**, 48–60 (2017). https://doi.org/10.1002/fes3.108
8. N.S. Sulaiman, K. Rovina, V.M. Joseph, Classification, extraction and current analytical approaches for detection of pesticides in various food products. J. Consum. Prot. Food Saf. **14**, 209–221 (2019). https://doi.org/10.1007/s00003-019-01242-4
9. Y. Abubakar, Y.Y. Deeni, Z. Fatima, O. Olubukola, Pesticides, history, and classification, in *Natural Remedies for Pest, Disease and Weed Control* (Academic Press, 2020), pp. 29–42. https://doi.org/10.1016/B978-0-12-819304-4.00003-8
10. D. Capoferri, F. Piazzolla, S. Castellano, M. Longhi, Affinity sensing strategies for the detection of pesticides in food. Foods **7**, 148 (2018). https://doi.org/10.3390/foods7090148
11. V. Dhananjayan, S. Jayakumar, B. Ravichandran, Conventional methods of pesticide application in agricultural field and fate of the pesticides in the environment and human health, in *Controlled Release of Pesticides for Sustainable Agriculture* (Academic Press, 2020), pp. 1–39. https://doi.org/10.1007/978-3-030-23396-9_1
12. T. Talaviya, D. Shah, N. Patel, H. Yagnik, M. Shah, Implementation of artificial intelligence in agriculture for optimisation of irrigation and application of pesticides and herbicides. Artif. Intell. Agric. **4**, 58–73 (2020). https://doi.org/10.1016/j.aiia.2020.04.002
13. R. Carson, *Silent Spring* (Houghton Mifflin Harcourt, Boston, MA, USA, 2002). https://doi.org/10.4324/9781315698724
14. T. Martinović, U. Andjelković, M.S. Gajdošik, D. Rešetar, D. Josić, Foodborne pathogens and their toxins. J. Proteomics **147**, 226–235 (2016). https://doi.org/10.1016/j.jprot.2016.04.029
15. R.V. Tauxe, Emerging foodborne pathogens. Int. J. Food Microbiol. **78**, 31–41 (2002). https://doi.org/10.1016/S0168-1605(02)00232-5
16. D. Yan, Y. Zhang, L. Liu, H. Yan, Pesticide exposure and risk of Alzheimer's disease: a systematic review and meta-analysis. Sci. Rep. **6**, 32222 (2016). https://doi.org/10.1038/srep32222
17. N.K. Rajawat, K. Bhardwaj, N. Mathur, Risk of Parkinson disease associated with pesticide exposure and protection by probiotics. Mater. Today Proc. (2023). https://doi.org/10.1016/j.matpr.2022.12.153
18. A. Kumar, S.L. Margekar, P. Margekar, V. Margekar, Recent advances in management of organophosphate & carbamate poisoning. Int. J. Med. Sci. **2018**, 7 (2018). https://doi.org/10.1016/j.injms.2018.06.007
19. M. Bagchi, S. Ghosh, D. Bagchi, E. Hassoun, S.J. Stohs, Protective effects of lazaroid U74389F (16-Desmethyl Tirilazad) on endrin-induced lipid peroxidation and DNA damage in brain and liver and regional distribution of catalase activity in rat brain. Free Radic. Biol. Med. **19**, 993–1000 (1995). https://doi.org/10.1016/0891-5849(95)00088-F

20. A. Oliva, A. Giami, L. Multigner, Environmental agents and erectile dysfunction: a study in a consulting population. Int. J. Impot. Res. **14**(3), S9–S15 (2002). https://doi.org/10.1002/j.1939-4640.2002.tb02277.x
21. J.K. Wipff, Gene flow in turf and forage grasses (Poaceae). Plants **1**, 143–161 (2002). https://doi.org/10.3390/plants1010143
22. V.F. Garry, M.E. Harkins, L.L. Erickson, L.K. Long-Simpson, S.E. Holland, B.L. Burroughs, Birth defects, season of conception, and sex of children born to pesticide applicators living in the Red River Valley of Minnesota, USA. Environ. Health Perspect. **110**(3), 441–449 (2002). https://doi.org/10.1289/ehp.02110s3441
23. T. Mattila, T. Santonen, H.R. Andersen, A. Katsonouri, T. Szigeti, M. Uhl, W. Wąsowicz, R. Lange, B. Bocca, F. Ruggieri, M. Kolossa-Gehring, D.A. Sarigiannis, H. Tolonen, Scoping review—the association between asthma and environmental chemicals. Int. J. Environ. Res. Public Health **18**, 1323 (2021). https://doi.org/10.3390/ijerph18031323
24. J.Y. Islam, J. Hoppin, A.M. Mora, M.E. Soto-Martinez, L. Córdoba Gamboa, J.E. Peñaloza Castañeda, B. Reich, C. Lindh, B. van Wendel de Joode, Respiratory and allergic outcomes among 5-year-old children exposed to pesticides. Thorax (2021). https://doi.org/10.1136/thoraxjnl-2021-218068
25. V. Kumar, N. Sharma, P. Sharma, R. Pasrija, K. Kaur, M. Umesh, B. Thazeem, Toxicity analysis of endocrine disrupting pesticides on non-target organisms: a critical analysis on toxicity mechanisms. Toxicol. Appl. Pharmacol. (2023). https://doi.org/10.1016/j.taap.2023.116623
26. S. Wahab, M.Y. Alshahrani, M.F. Ahmad, H. Abbas, Current trends and future perspectives of nanomedicine for the management of colon cancer. Eur. J. Pharmacol. (2021). https://doi.org/10.1016/j.ejphar.2021.174464
27. P. Rebouillat, R. Vidal, J.-P. Cravedi, B. Taupier-Letage, L. Debrauwer, L. Gamet-Payrastre, M. Touvier, M. Deschasaux-Tanguy, P. Latino-Martel, S. Hercberg, Prospective association between dietary pesticide exposure profiles and postmenopausal breast-cancer risk in the NutriNet-Santé Cohort. Int. J. Epidemiol. **50**, 1184–1198 (2021). https://doi.org/10.1093/ije/dyab015
28. L.S. Engel, E.C. Zabor, J. Satagopan, A. Widell, N. Rothman, T.R. O'Brien, M. Zhang, S.K. Van Den Eeden, T.K. Grimsrud, Prediagnostic serum organochlorine insecticide concentrations and primary liver cancer: a case-control study nested within two prospective cohorts. Int. J. Cancer (2019). https://doi.org/10.1002/ijc.32175
29. K.R. Greenop, S. Peters, H.D. Bailey, L. Fritschi, J. Attia, R.J. Scott, D.C. Glass, E. Milne, L.J. Ashton, B.K. Armstrong et al., Exposure to pesticides and the risk of childhood brain tumors. Cancer Causes Control **24**, 1269–1278 (2013). https://doi.org/10.1007/s10552-013-0205-1
30. C.A. Lucchesi, D.M. Vasilatis, S. Mantrala, T. Chandrasekar, M. Mudryj, P.M. Ghosh, Pesticides and bladder cancer: mechanisms leading to anti-cancer drug chemoresistance and new chemosensitization strategies. Int. J. Mol. Sci. **24**, 11395 (2023). https://doi.org/10.3390/ijms241411395
31. A. Rafeeinia, G. Asadikaram, V. Moazed, M. Karimi Darabi, Organochlorine pesticides may induce leukemia by methylation of CDKN2B and MGMT promoters and histone modifications. Gene (2022). https://doi.org/10.1016/j.gene.2022.146976
32. D.D. Weisenburger, A review and update with perspective of evidence that the herbicide glyphosate (roundup) is a cause of Non-Hodgkin Lymphoma. Clin. Lymphoma Myeloma Leuk. (2021). https://doi.org/10.1016/j.clml.2021.04.009
33. Codex Alimentarius Commission, *Pesticide Residues in Food and Feed*. Available online: http://www.fao.org/fao-who-codexalimentarius/codex-texts/dbs/pestres/en
34. World Health Organization (WHO), *Joint FAO/WHO Meeting on Pesticide Residues (JMPR)*. Available online: https://www.who.int/foodsafety/areas_work/chemical-risks/jmpr/en/
35. European Union, *Regulation (EC) No 396/2005 on Maximum Residue Levels of Pesticides* (2005). Available online: https://eur-lex.europa.eu/legal-content/EN/TXT/?uri=CELEX%3A32005R0396
36. U.S. Environmental Protection Agency, *Pesticide Residue Tolerances* (2020). Available online: https://www.epa.gov/pesticide-tolerances

37. Food Safety and Standards Authority of India (FSSAI), *Food Safety and Standards (Contaminants, Toxins, and Residues) Regulations* (2011). Available online: https://www.fssai.gov.in/cms/contaminants-toxins-residues.php
38. Government of India, *The Food Safety and Standards Act* (2006). Available online: https://www.fssai.gov.in/cms/food-safety-and-standards-act-2006.php
39. Food Safety and Standards Authority of India (FSSAI), *Manual of Methods of Analysis of Foods: Pesticide Residues* (2015). Available online: https://fssai.gov.in/cms/manuals-of-methods-of-analysis-for-various-food-products.php
40. D. Hamilton, S. Crossley, *Pesticide Residues in Food and Drinking Water: Human Exposure and Risks* (Wiley, Hoboken, NJ, USA, 2004)
41. C.F. Poole, *Gas Chromatography* (Elsevier, Amsterdam, The Netherlands, 2012)
42. S. Hassani, S. Momtaz, F. Vakhshiteh, A.S. Maghsoudi, M.R. Ganjali, P. Norouzi, M. Abdollahi, Biosensors and their applications in detection of organophosphorus pesticides in the environment. Arch. Toxicol. **91**, 109–130 (2017). https://doi.org/10.1007/s00204-016-1875-8
43. Jeanette, *Immunoassay and Other Bioanalytical Techniques* (Informa, 2016). https://doi.org/10.1201/9781420020694
44. D.A. Skoog, F.J. Holler, S.R. Crouch, *Principles of Instrumental Analysis* (Cengage Learning, Boston, MA, USA, 2017)
45. J.M. Van Emon, *Immunoassay and Other Bioanalytical Techniques* (CRC Press, Boca Raton, FL, USA, 2007)
46. U. Jain, K. Saxena, V. Hooda, S. Balayan, A.P. Singh, M. Tikadar, N. Chauhan, Emerging vistas on pesticides detection based on electrochemical biosensors—an update. Food Chem. **371**, 131126 (2022). https://doi.org/10.1016/j.foodchem.2021.131126
47. International Organization for Standardization, *ISO 17025: General Requirements for the Competence of Testing and Calibration Laboratories* (International Organization for Standardization, Geneva, Switzerland, 2019)
48. M. Thompson, S.L.R. Ellison, R. Wood, Harmonized guidelines for single-laboratory validation of methods of analysis. Pure Appl. Chem. **74**, 835–855 (2002)
49. J.K. Taylor, *Quality Assurance of Chemical Measurements* (CRC Press, Boca Raton, FL, USA, 1987)
50. S.O. Aiyedun, S.J. Kim, L.Q. Luu, S. Charlebois, H.K. Bektas, J. Chen, Foodborne outbreaks of microbial infection from fresh produce in Europe and North America: a systematic review of data from this millennium. Int. J. Food Sci. Technol. **56**, 2215–2223 (2021). https://doi.org/10.1111/ijfs.14884
51. F. Yeni, S. Yavaş, H. Alpas, Y. Soyer, Most common foodborne pathogens and mycotoxins on fresh produce: a review of recent outbreaks. Crit. Rev. Food Sci. Nutr. **56**, 1532–1544 (2016). https://doi.org/10.1080/10408398.2013.777021
52. L. Chen, J. Yang, W. Luo, S. Zhang, L. Ma, P. Wei, Surveillance for foodborne disease outbreaks in Zhejiang Province, China, 2015–2020. BMC Public Health **22**, 135 (2022). https://doi.org/10.1186/s12889-022-12568-4
53. A. Bisht, S. Sachdev, N. Chopra, H. Vashist, A surveillance of food borne disease outbreaks in India: 2009–2018. Food Control **121**, 107630 (2021). https://doi.org/10.1016/j.foodcont.2020.107630
54. M.P. Kabiraz, M. Kamruzzaman, M.A. Hossain, M.M. Rahman, Conventional and advanced detection techniques of foodborne pathogens: a comprehensive review. Heliyon e13804 (2023). https://doi.org/10.1016/j.heliyon.2023.e15482
55. Y. Wang, K. Jia, J. Lin, Optical biosensors for the detection of foodborne pathogens: recent development and future prospects. TrAC Trends Anal. Chem. 117785 (2024). https://doi.org/10.1016/j.trac.2024.117785
56. H. Hu, S. Chen, Z. Wang, Y. Zhu, X. Sun, Z. Zhao, J. Xiao, L. Zhou, A robust and facile colorimetric aptasensor for the detection of *Salmonella typhimurium* based on the regulation of Fe_3O_4@Cu@PCPy Yolk-Shell Nanozyme activity. Anal. Chim. Acta **1276**, 341618 (2023). https://doi.org/10.1016/j.aca.2023.341618

57. F. Xiao, Y. Xu, W. Zhou, Q. Hu, D. Zhang, S. Du, Z. Zhu, C. Shen, Smartphone-assisted biosensor based on broom-like bacteria-specific magnetic enrichment platform for colorimetric detection of *Listeria monocytogenes*. J. Hazard. Mater. **459**, 132250 (2023). https://doi.org/10.1016/j.jhazmat.2023.132250
58. B. Zhang, X. Feng, Y. Yu, Y. Zhang, H. Jiang, J. Wang, Polyethyleneimine-interlayered silica-core quantum dot-shell nanocomposites for sensitive detection of *Salmonella typhimurium* via a lateral flow immunoassay. RSC Adv. **10**, 2483–2489 (2020). https://doi.org/10.1039/C9RA09252H
59. Y. Ding, Z. Ma, P. Wang, F. Zhang, J. Xu, W. Chen, L. Li, Quantum dot-labeled phage-encoded RBP 55 as a fluorescent nanoprobe for sensitive and specific detection of *Salmonella* in food matrices. Food Chem. **428**, 136724 (2023). https://doi.org/10.1016/j.foodchem.2023.136724
60. D. Bhandari, F.C. Chen, R.C. Bridgman, Detection of *Salmonella typhimurium* in romaine lettuce using a surface Plasmon resonance biosensor. Biosensors **9**, 94 (2019). https://doi.org/10.3390/bios9030094
61. H.V. Raghu, N. Kumar, Rapid detection of *Listeria monocytogenes* in milk by surface Plasmon resonance using wheat germ agglutinin. Food Anal. Methods **13**, 982–991 (2020). https://doi.org/10.1007/s12161-020-01717-3
62. S.-M. You, G. Kim, S.R. Kim, S.Y. Oh, J. Lim, S.-K. Cho, M. Kang, K. Shin, S.H. Lee, S.M. Yoo, S.Y. Lee, Gold nanoparticle-coated starch magnetic beads for the separation, concentration, and SERS-based detection of *E. coli* O157. ACS Appl. Mater. Interfaces **12**, 18292–18300 (2020). https://doi.org/10.1021/acsami.0c00418
63. F. Mi, X. Wang, W. Zhang, H. Fang, Y. Wang, G. Chen, C. Zhang, X. Huang, F. Yang, J. Nie, Integration of three non-interfering SERS probes combined with ConA-functionalized magnetic nanoparticles for extraction and detection of multiple foodborne pathogens. Microchim. Acta **190**, 103 (2023). https://doi.org/10.1007/s00604-023-05676-4

Spectroscopic Techniques for Detecting Food Adulterants

P. J. Dihingia and S. Rai

Abstract In this chapter, the principles and techniques of different optical probes used in the detection and quantification of food adulterants have been described with emphasis on infrared (IR) spectroscopy. The IR spectroscopy provides rapid detection and quantification of a variety of adulterants in common food items such as milk, honey, cooking oil, spices, flour, etc. For example, the distinctive 1140 cm^{-1} peak in FTIR spectrum of turmeric powder indicates the presence of metanil, a common adulterant.

Keywords Optical probes · Food adulterants · IR spectroscopy

1 Introduction

Optical spectroscopic methods, nowadays, are fast replacing the traditional methods of detecting food adulterants. The optical methods include UV–Vis spectroscopy, IR spectroscopy, Raman spectroscopy, LIBS (Laser Induced Breakdown Spectroscopy), SPR (Surface Plasmon Resonance), Hyperspectral imaging, etc. Different miniature probes and sensors have been developed on the basis of these methods. This has led to the portability of these probes and facilitates on-site examination of food adulteration or contamination. The old techniques such as chromatography and immunoassays involve a lot of time and labor. The optical sensors, on the other hand, provide fast, accurate and cost-effective detection of various food adulterants. The non-destructive nature of these probes ensures that the physical and chemical properties of the samples do not change during analysis. Moreover, the quantity of the sample required for these methods is very less compared to others and hence unnecessary wastage of some valuable food items can be prevented. Accordingly, this chapter overviews the

P. J. Dihingia (✉)
Department of Physics, Dibru College, Dibrugarh, Assam, India
e-mail: pranab.dihingia@gmail.com

S. Rai
Department of Physics, Mizoram University, Aizawl, Mizoram, India

R. Biswas (ed.), *Optical Techniques for Assessing Food Adulterants*, Progress in Optical Science and Photonics 35, https://doi.org/10.1007/978-981-96-8142-6_3

application of spectroscopic techniques for assessing food adulteration with special emphasis on IR spectroscopy.

1.1 UV–Vis Spectroscopy

It involves passing ultraviolet (UV) and visible (Vis) light through a transparent or semi-transparent sample and measuring the "absorbance (A)" or "optical density (O.D.)" of the sample with the help of a detector. The instrument is thus called UV–Vis spectrophotometer. Today's spectrophotometers are computer controlled; strong software assists in scanning the sample over a selected wavelength range and retrieving the absorbance data for future use. A typical plot between wavelength and absorbance is known as absorption spectrum. The spectrum obtained contains some peaks which are the characteristics of a particular sample. A good UV–Vis spectrophotometer provides a wide range of wavelengths for scanning the samples, starting from 190 to 800 nm. The UV–Vis spectroscopy is used in detecting a variety of adulterants of common food items such as melamine in milk, palm oil in olive oil, etc. Also, the technique is helpful in analyzing milk fat, evaluation of edible oil, estimating vitamin C levels in fruits and vegetables, etc. [1].

The principle of UV–Vis spectroscopy is relatively simple. A photon of specific wavelength, on hitting a molecule of the sample, excites it from a low energy state to a high energy state. The electronic transition takes place by absorbing the energy of the photon which is equal to the difference of energy between the two levels. This absorbed light is detected by the detector and is further processed and recorded with the help of software. The Beer-Lambert law gives the absorbance of the sample as

$$A = \varepsilon l c$$

where ε is the molar absorptivity of the sample, l is the optical path length and c is the concentration of the sample. The absorbance is the amount of light absorbed by the material of the sample. The transmittance (T) is just the opposite of absorbance, i.e. it gives the amount of light transmitted through sample after absorption. It is given by

$$T = \frac{I}{I_o}$$

where I is the intensity of light transmitted through the sample and I_o is the intensity of light transmitted through blank (i.e., when there is no sample in the sample holder). A and T are related as

$$A = -\log(T)$$

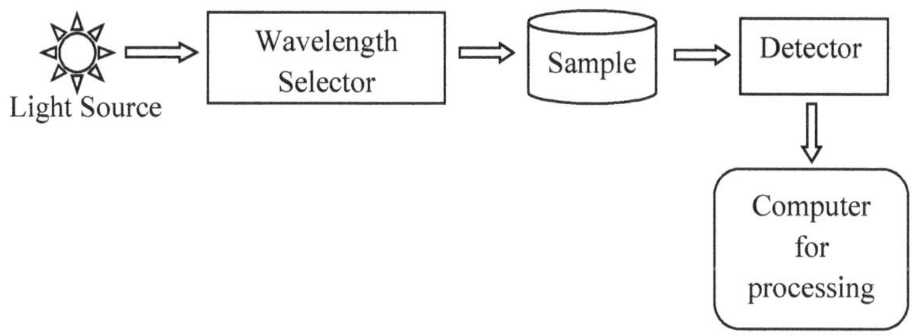

Fig. 1 The block diagram for UV–Vis spectroscopy

A typical UV–Vis spectrophotometer consists of a light source, a wavelength selector, a sample holder and a detector as its main components (Fig. 1).

The light source used in a UV–Vis spectrophotometer is generally high intensity xenon lamp, but there can be two sources used for the same-one is the deuterium lamp for UV region and another is a halogen lamp for the visible region. The wavelength selector allows us to select a desired range of wavelengths to scan the samples. The sample holder is popularly known as "cuvette", which is a rectangular shaped glass sample holder. The two opposite faces of a cuvette are opaque and the other two are optically flat. The detector usually consists of a photodiode array which coverts light signals into corresponding electrical signals.

1.2 IR Spectroscopy

The infrared (IR) spectroscopy utilizes the IR region of the electromagnetic spectrum for probing into the molecular structure of materials. Hence, it can be used for identification of different kinds of materials. Although, the IR region is divided into three parts, viz. near IR, mid IR and far IR, yet the wavelengths between 2.5 and 16 μm are generally useful for detecting specific organic compounds in a sample.

When the frequency of the IR radiation matches with that of particular bond or group of a molecule, then the molecule absorbs that radiation. This is a resonant process influenced by the potential energy of the molecule, the molecular mass and the coupling between different electronic states. In fact, these resonant frequencies determine the presence of certain chemicals in a food material. In an IR spectrometer, the infrared light is sent through the sample and the IR spectrum is obtained by plotting the measured intensity of the transmitted light versus frequency (in cm^{-1}). Nowadays, the Fourier-Transform Infrared Spectroscopy (FTIR) has become more popular than traditional IR spectroscopy because it is faster and more sensitive.

In an IR spectrometer, a prism or grating separates infrared light into its component wavelengths and the intensity of each wavelength is measured by a detector after passing through the sample (Fig. 2). But in an FTIR spectrometer, simultaneous data

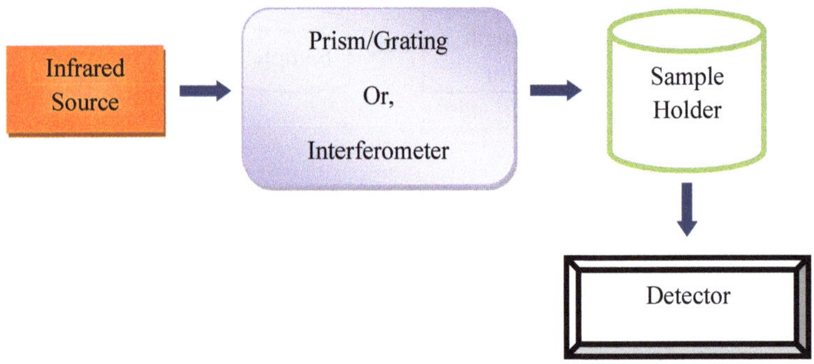

Fig. 2 The block diagram of an IR spectrometer

collection over a broad range of IR wavelengths is done by an interferometer. These data are then Fourier transformed using an algorithm to get the IR spectrum.

1.3 Raman Spectroscopy

If a substance is illuminated with a light source such as laser, then the scattered light off the sample carries a lot of information about its chemical structure and hence this information can be used to identify the substance. This is the working principle of Raman spectroscopy, named after Nobel Prize winning Indian physicist C.V. Raman who along with his associates discovered Raman scattering in 1928.

The light scattered by a molecule of a substance can be either elastic or inelastic. In elastic or Rayleigh scattering, the scattered light has the wavelength or colour as that of the incident light. In inelastic or Raman scattering, the scattered photons have wavelengths different from that of incident photons. Some scattered photons possess wavelengths greater than that of striking photons and we get Stokes' lines in the Raman spectrum corresponding to those scattered photons. On the other hand, the scattered photons possessing wavelengths less than that of incident photons give Anti-Stokes' lines in the Raman spectrum (Fig. 3). The Raman shift ($\Delta\overline{\upsilon}$) is determined by the frequency difference between the original photon and the scattered photons. It is usually expressed in cm^{-1}. A typical Raman spectrum is a plot between Raman shift and intensity of the scattered light.

The molecules in the sample absorb that amount of incident light energy which is equal to that of the natural frequency of vibration of the bonds in molecules. Since different molecules have different bonds, therefore, the amount of energy absorbed from the probing light will be not be the same for all the molecules. Hence, the scattered light will be a mix of photons having wavelengths both greater and less than the incident light. The Raman spectrum analysis thus becomes an invaluable tool in finding out traces of contamination or foreign material in a pure substance.

Fig. 3 The schematic diagram of Raman scattering

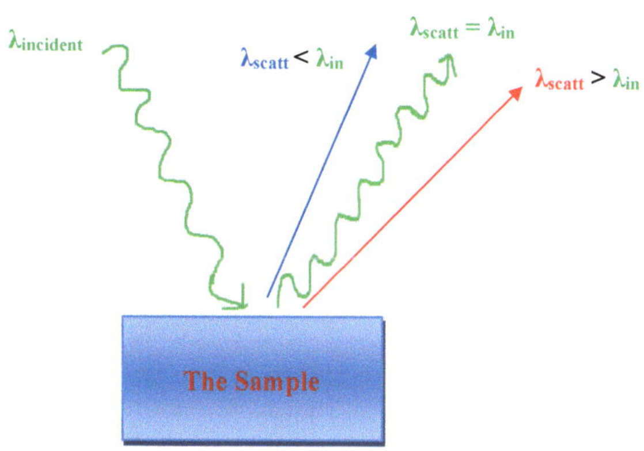

This spectroscopic technique is particularly helpful in the detection of high fructose corn syrup in honey [2], Sudan dye in chili powders [3], etc.

The basic components of a Raman spectrometer are: excitation source, sampling apparatus and detector. Modern Raman spectrometers contain a laser as excitation source and a fibre optic probe or microscope as sampling apparatus. The wavelength of the laser is determined by the fact that a strong Raman spectrum is obtained only when relatively shorter wavelengths are used. Moreover, the organic molecules exhibit fluorescence when short wavelength photons are used to excite them.

1.4 LIBS

The Laser Induced Breakdown Spectroscopy (LIBS in short) is an in-situ atomic spectroscopic technique used for detecting major and trace elements in a solid or liquid sample. In LIBS, a micro-plasma is created on the sample surface using a short laser pulse. The laser causes ablation of small amount of mass off from surface of the sample. The short-lived high-temperature plasma is the result of interaction of this ablated mass with trailing portion of the laser pulse (Fig. 4). This plasma contains electrons, ions and atoms in higher energy states. As the plasma cools off, the electrons come down from higher excited electronic states to their natural ground states. In the process, photons are emitted whose energy is equal to the difference between the excited states and the ground state. These are recorded as discrete spectral peaks in the LIBS emission spectra. Now, the chemical composition of the sample can be determined by analyzing different peaks. The LIBS emission peak intensities are used to quantify the concentration of the elements in the sample. It can detect heavy metallic elements in the low-PPM range. LIBS is especially helpful in the detection of sweetener syrups in honey [4] and their subsequent quantification. Also, LIBS combined with chemometric methods can detect adulteration of pistachio with green pea and spinach samples [5]. LIBS is used for identification of milk fraud as well [6].

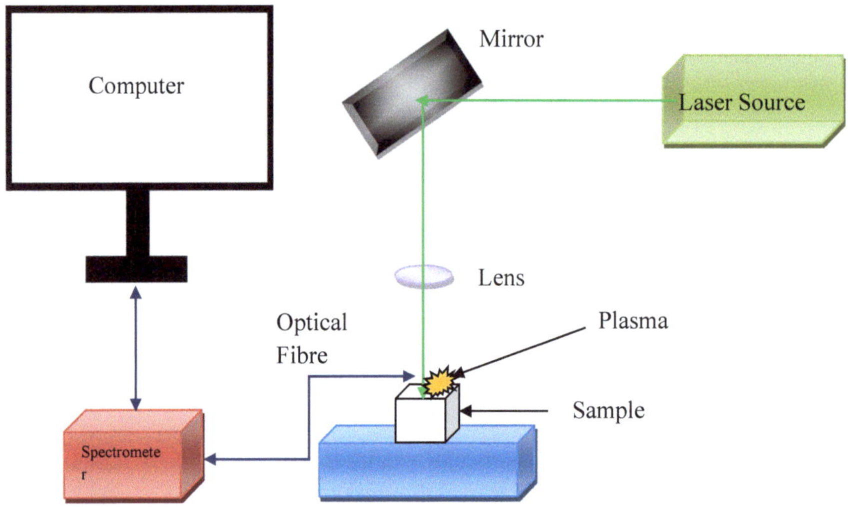

Fig. 4 The schematic diagram of a LIBS system

1.5 SPR

The Surface Plasmon Resonance (SPR in short) is a spectroscopic technique used for measuring molecular interactions in real-time. It occurs when plane-polarized light becomes incident on a metal film at the interface separating two media. The angle of incidence should be such that it fulfills the condition of total internal reflection. This technique involves excitation and detection of "surface plasmons", which are the collective oscillations of free electrons. Here, the light is focused onto a metal film through a glass prism and the subsequent reflection is detected. At the critical angle of incidence, the plasmons resonate with the polarized light and as a result the polarized light is absorbed by the sample on the metal film. The sample consists of a "ligand" to which the "analyte" is bounded and immobilized on the metal film (the sensor chip) (Fig. 5). Due to this absorption, a dark line is obtained in the reflected pattern. The unit of reflected pattern recorded by the instrument is RU (Resonance Units), which is proportional to the mass on the film or the number of molecules bound to the surface. The SPR sensors are used in the detection of allergens, pathogens, pesticides and various other toxic chemicals in food items [7–9].

1.6 Hyperspectral Imaging

The Hyperspectral Imaging (HSI) is non-invasive and non-destructive technology combining both spectroscopy and imaging. It enables detail identification and quantification of materials, which are much superior to other imaging systems. In HSI,

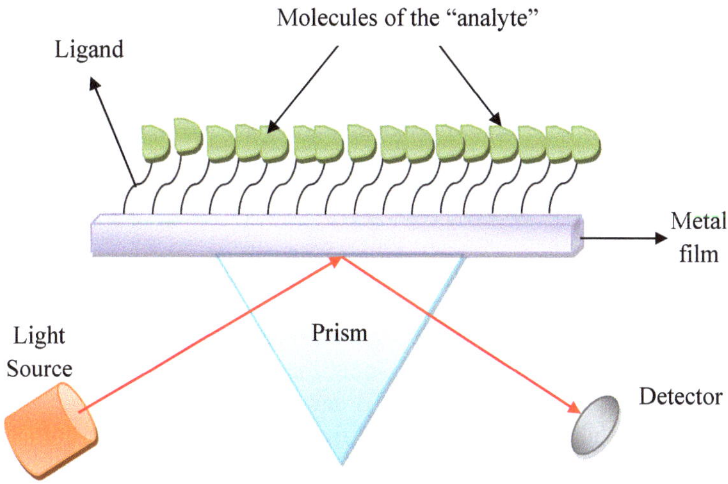

Fig. 5 The schematic representation of SPR technique

a spectrum is generated for each "pixel" of an image by collecting and processing information across the entire electromagnetic spectrum. Thus, the identity of materials can be found out from their specific spectral patterns. The typical range of observation with HSI includes the band 250–15,000 nm and thermal infrared.

HSI uses an imaging spectrometer known as hyperspectral camera to gather information about an object. The hyperspectral camera collects the reflected light from the object and separates it into individual wavelengths. Since each pixel of an image produces a unique spectrum, therefore collecting those spectrums gives a two-dimensional hyperspectral image of the object. The utility of HSI in detecting different compounds lies in the fact that each material interacts with light differently and hence their reflected spectra also vary. HSI can provide both spectral and spatial information about an object's physical and chemical properties.

The various uses of HSI are—environmental monitoring, mineral exploration, quality control, waste management, agriculture, military surveillance, etc. Especially, the HSI-assisted in-line quality control systems in food industries enable detection and identification of contaminants and quantity of salt, sugar, fat, or moisture in a particular food product.

2 Infrared Spectroscopic Based Methods for Detecting Food Adulterants

The Infrared (IR) spectroscopy provides a very convenient method for quick and easy detection of food adulterants [10]. In combination with chemometric techniques, it provides an effective tool for detection of adulterants in common food items such as milk, honey, cooking oil, etc., as well as high value food products such as wine, dietary supplements, etc. [11]. Here, we give a detail review of application of IR spectroscopy for detecting adulterants in different food items.

2.1 Adulterants in Milk

The adulteration of milk is a nasty practice for a section of people of our society who want to earn huge profit in a short time, but do not insist on investing more on production and quality. It's just not a practice of individuals running small production facilities or milkman going door-to-door for selling milk, but sometimes such activities have also been reported from huge and reputed companies too. The biggest scandal in milk adulteration was reported from China in 2008 [12]. In that year, sixteen infants in China's Gansu province had been diagnosed with kidney stones and later it was found that all of them were fed powdered milk containing toxic compound "melamine". The milk was allegedly produced by Sanlu Group, but later many more dairy farms were found to be involved in the scandal. Lu et al. (2009) proposed a method for detecting melamine in milk [13] using near infrared spectroscopy (NIR) combined with least squares-support vector machine (LS-SVM). The use of single bounce attenuated total reflectance (SB-ATR) Fourier transform infrared spectroscopy (FTIR) for detection and quantification of melamine in dairy milk was shown by Sana et al. [14] in 2013. Chen et al. (2017) also reported detection of melamine adulteration in milk using NIR spectroscopy and one-class partial least squares method [15]. Another common adulterant in milk is urea, which is responsible for serious stomach problems. Jha et al. (2015) evaluated the attenuated total reflectance (ATR)-Fourier transform infrared spectroscopy (FTIR) for rapid detection and quantification of added urea in milk [16]. They utilized the wavelength range 4000–700 cm^{-1} for the purpose. Formalin is added as an adulterant to cow milk for increasing shelf life, which is detrimental for kidney and liver. Mabood et al. (2016) utilized NIR spectroscopy combined with multivariate analysis to detect and quantify the level of formalin in cow milk [17]. The most common preservative for milk is benzoic acid. But, for sensitive persons, it may cause asthma, urticaria, metabolic acidosis and convulsions even at low doses [18]. Cattaneo and Holroyd (2013) reviewed the importance of NIR spectroscopic method for identification and quantification of various adulterants in milk including benzoic acid [19].

2.2 Adulterants in Honey

Sugar is the most frequently used adulterant in honey. Sivakesava and Irudayaraj (2001) used FTIR spectroscopy with attenuated total reflection (ATR) sampling accessory for determination of cane medium invert sugar in three different varieties of honey [20]. They developed predictive models to classify cane sugar-adulterated honey samples using discriminant analysis. Cozzolino et al. (2011) also reviewed IR spectroscopic methods for detecting adulterants and residues in honey [21].

2.3 Adulterants in Mustard Oil

Mustard oil is the most used cooking oil in western, northern and eastern India. There are a large number of brands of mustard oil available in Indian markets. To get the maximum health benefits, we must ensure that the mustard oil is adulterant-free. The most common adulterant in mustard oil is argemone oil. Jamwal et al. (2020) analyzed the potential of ATR-FTIR spectroscopy coupled with multivariate chemometrics for the detection of argemone oil in mustard oil [22]. They utilized the frequency ranges 3050–2750 and 1800–500 cm^{-1} for the separation of mustard oil from argemone oil. Another common adulterant of mustard oil is linseed oil. ATR-FTIR spectroscopy can also be applied for the detection of mustard oil adulteration with linseed oil [23]. The butter yellow adulteration in mustard oil can also be detected using NIR spectra [24].

2.4 Adulterants in Olive Oil

Olive oil is a healthier alternative for those who do not want to use mustard oil as cooking oil. It is rich in mono-unsaturated fats and large amounts of antioxidants. Moreover, it has strong anti-inflammatory properties. But often this oil is adulterated with cheap palm oil. FTIR and FT-NIR spectroscopy have already been used for the analysis of extra-virgin olive oil (EVOO) adulterated with palm oil [25, 26]. Lai et al. (1995) reported quantitative analysis of adulterants in EVOO such as refined olive oil and walnut oil using IR spectroscopy [27]. The detection of adulteration of EVOO with other edible oils such as corn oil, sunflower oil, cottonseed oil and rapeseed oil is also made possible with chemometric analysis of mid-infrared (MIR) spectra [28].

2.5 Adulterants in Turmeric Powder

Turmeric is an essential ingredient in every Indian and many South-Asian dishes. It has anti-inflammatory and antioxidant properties, which are beneficial for human body. The growing demand of turmeric powder has lead to its adulteration with metanil yellow, which is an azo dye. Metanil yellow has a damaging effect to eyes, may cause infertility and skin irritation in human. Dia et al. (2024) have reported the detection of metanil yellow in turmeric powder using FTIR spectroscopy combined with chemometrics OPLS-DA and PLS [29]. The distinctive 1140 cm^{-1} peak can be assigned as due to metanil yellow. IR spectroscopy can also be applied for detection of other adulterants in turmeric powder such as Sudan Red, starch, etc. [30].

2.6 Adulterants in Red Chili Powder

The preparation of most of the Indian dishes remains incomplete without the addition of red chili powder. It enhances the flavor and visual appearance of food items. Unfortunately, red chili powder from many famous brands has been found to be adulterated with chemicals such as Sudan I dye, etc. NIR spectroscopy is found to be very useful in detecting Sudan dye in red chili powder [3]. Avocado pear seeds and cola nuts are some other adulterants found in chili pepper powder, which can be also be detected with NIR spectroscopy combined with chemometrics [31].

2.7 Adulterants in Wine

The common adulterants in wine are toxic alcohols such as methanol, dyes and ammonium nitrate. In 2006, a study revealed that in India 64% of the liquors were adulterated with methanol, which is so dangerous that consuming only 30 ml of it can be fatal. The near infra-red (NIR) spectroscopy with attenuated total reflectance (ATR) is a reliable tool for detection and quantification of adulterants in wine and other liquors [32].

2.8 Adulterants in Coffee and Tea

Tea is a popular drink in both India and China. But some traders make undue profit by adding essence to substandard tea. The characteristic aroma is a signature of high-quality tea. Although traditional physical and chemical methods can detect volatile organic compounds such as essence in tea, they take a lot of time for the same. The Fourier transformed infrared (FTIR) spectroscopy can rapidly detect the presence of essence in various tea samples [33]. On the other hand, the NIR spectroscopy can detect the presence of barley and corn in coffee [34, 35].

2.9 Adulterants in Wheat Flour

The wheat flour is considered as the most desired cereal flour and it is expensive too for manufacturing many food products. But, this wheat flour is often found to be adulterated with barley flour which is less expensive cereal flour. The FTIR spectroscopy with chemometrics can be deployed for rapid detection of barley flour in wheat flour [36].

3 Future of IR Spectroscopy in Detecting Food Adulterants

With the development of advanced semiconductor IR sensors, the IR spectroscopes such as FTIR spectrometers are going to be less bulky, making them portable and very handy. This will ultimately lead to detection of adulterants in food items on-site, i.e., whenever and wherever such adulteration is suspected, the detection should be possible at the place itself.

Remarks

Out of many spectroscopic techniques, the IR spectroscopy is particularly helpful in providing information about the presence or absence of specific functional groups. These functional groups are the "signatures" of various adulterants in food items. So, by identifying the functional groups present in a sample food material, we can detect the presence of one or more adulterants in it.

References

1. F. Haque, S.Y. Bubli, M.S. Khan, *Techniques to Measure Food Safety and Quality, Chapter 8* (2021)
2. S. Li, Y. Shan, X. Zhu, X. Zhang, G. Ling, J. Food Compos. Anal. **28**(1), 69 (2012)
3. S.A. Haughey, P. Galvin-King, Y. Ho, S.E.J. Bell, C.T. Elliott, Food Control **48**, 75 (2015)
4. M.G. Nespeca, A.L. Vieira, D.S. Junior, J.A.G. Neto, E.C. Ferreira, Food Chem. **311**, 125886 (2020)
5. B. Sezer, H. Apaydin, G. Bilge, Ismail H. Boyaci, J. Sci. Food Agric. **99**(5), 2236 (2018)
6. B. Sezer, S. Durna, G. Bilge, A. Berkkan, A. Yetisemiyen, I.H. Boyaci, Int. Dairy J. **81**, 1 (2018)
7. M.T.I. Ansari, S.K. Raghuwanshi, S. Kumar, TNB. **22**(4), 978 (2023)
8. N. Ravindran, C.A. Mamathi, N.S Thirunavookarasu, C.K. Sunil, Crit. Rev. Food Sci. Nutr. **63**(8), 1055 (2023)
9. J. Zhou, Q. Qi, C. Wang, Y. Qian, G. Liu, Y. Wang, L. Fu, Biosens. Bioelectron. **142**(1), 111449 (2019)
10. R. Valand, S. Tanna, G. Lawson, L. Bengtström, Food Addit. Contam. Part A **37**, 19 (2020)
11. E. Mendes, N. Duarte, Foods **10**(2), 477 (2021)
12. Y. Huang, The 2008 Milk Scandal Revisited, Forbes Asia (2014)
13. C. Lu, B. Xiang, G. Hao, J. Xu, Z. Wang, C. Chen, J. Near Infrared Spectrosc. **17**, 59 (2009)
14. S. Jawaid, F.N. Talpur, S.T.H. Sherazi, S.M. Nizamani, A.A. Khaskheli, Food Chem. **141**(3), 3066 (2013)
15. H. Chen, C. Tan, Z. Lin, T. Wu, Spectrochim. Acta A Mol. Biomol. Spectrosc. **173**, 832 (2017)
16. S.N. Jha, P. Jaiswal, A. Borah, A.K. Gautam, N. Srivastava, Food Bioprocess Tech. **8**, 926 (2015)
17. F. Mabood, J. Hussain, M.O.O. Al Nabhani, S.A. Gilani, S. Farooq, Z. Naureen, F. Jabeen, M. Ahmed, Z. Hussain, A. Al-Harrasi, J. Adv. Dairy Res. **5**(01), 1 (2016)

18. P. Qi, H. Hong, X. Liang, D. Liu, Food Control **20**(4), 414 (2009)
19. T.M.P. Cattaneo, S.E. Holroyd, J. Near Infrared Spec. **21**(5), 341 (2013)
20. S. Sivakesava, J. Irudayaraj, J. Food Sci. **66**, 972 (2001)
21. D. Cozzolino, E. Corbella, H.E. Smyth, Appl. Spectrosc. Rev. **46**(7), 523 (2011)
22. R. Jamwal, S. Kumari, B. Balan, A.S. Dhaulaniya, S. Kelly, A. Cannavan, D.K. Singh, LWT **120**, 108945 (2020)
23. R. Jamwal, S. Kumari, S. Kelly, A. Cannavan, D.K. Singh, Vib. Spectrosc. **113**, 103226 (2021)
24. R. Amsaraj, S. Mutturi, Infrared Phys. Techn. **129**, 104543 (2023)
25. A. Rohman, Y.B. Che Man, Food Res. Int. **43**(3), 886 (2010)
26. H. Azizian, M.M. Mossoba, A.R. Fardin-Kia, P. Delmonte, S.R. Karunathilaka, J.K.G. Kramer, Lipids **50**(7), 705 (2015)
27. Y.W. Lai, E.K. Kemsley, R.H. Wilson, Food Chem. **53**(1), 95 (1995)
28. G. Gurdeniz, B. Ozen, Food Chem. **116**(2), 519 (2009)
29. S.M. Dia, A.R. Putri, L.A. Muchlashi, Indones. J. Chem. Anal. **7**(1), 64 (2024)
30. R. Khodabakhshian, M.R. Bayati, B. Emadi, Food Chem. **364**, 130406 (2021)
31. E.K. Essuman, E. Teye, L.K. Sam-Amoah, C.L.Y. Amuah, Infrared Phys. Techn. **135**, 104961 (2023)
32. A. Hencz, L.L.P. Nguyen, L. Baranyai, D. Albanese, Processes **10**, 95 (2022)
33. C. Yang et al., Food Chem. **423**, 136308 (2023)
34. H. Ebrahimi-Najafabadi et al., Talanta **99**, 175 (2012)
35. J.K. Winkler-Moser et al., J. Agric. Food Chem. **63**(49), 10662 (2015)
36. F.N. Arslan et al., J. Consum. Prot. Food S. **15**, 245 (2020)

Advancements in Fiber-Optic-Based SPR Biosensors for Detecting Food Adulteration

Dimpi Paul and Sibasish Dutta

Abstract Food safety is a scientific discipline that encompasses the handling, preparation and storage of food in such a way that can prevent food-borne illnesses and ensure no harm to the consumers. Also, food safety is a global challenge that has ignited growing concern among the consumers and food production industries as well. In the food production chain, practices related to food safety help maintain food quality and wholesomeness, promoting good health. To deal with such a situation one can move towards optical sensors that have the capacity to monitor the food health cost-effective and handy techniques. However, optical sensors have also emerged as a solution to overcome these constraints; that are named as biosensors. Biosensors offer more precise and rapid sensing techniques in comparison to rigorous lab techniques like chromatography and immunoassays. One of the most important advantages of biosensors is that it has the ability to detect food adulteration in a rapid, non-destructive and cost-effective method. Over the past few decades, there has been a significant interest in developing surface Plasmon resonance (SPR) and LSPR (localized surface plasmon resonance) sensors. These SPR sensors are used for detecting as well as monitoring various contaminants in foods, including pesticides, pathogens, allergens and toxic chemicals. Those ground-breaking sensors play a vital role in detecting various adulterants within the food matrix. Thus, in this particular chapter, we will explore the utilization of SPR and LSPR-based sensors and discuss the key challenges that one can encounter in their implementation.

Keywords Surface plasmon resonance (SPR) · Localized surface plasmon resonance (LSPR) · Chromatography

---

D. Paul (✉)
Department of Physics, Patharkandi College, Sribhumi, Assam 788724, India
e-mail: paulddimpi21@gmail.com

S. Dutta
Department of Physics, Pandit Deendayal Upadhyaya Adarsha Mahavidyalaya (PDUAM) Eraligool, Sribhumi, Assam 788723, India

R. Biswas (ed.), *Optical Techniques for Assessing Food Adulterants*, Progress in Optical Science and Photonics 35, https://doi.org/10.1007/978-981-96-8142-6_4

1 Introduction

We are in a century in which it is quiet difficult to get good quality food without any hassle. Now a day's, consumption of good quality food is a major concern among the people. This is the reason that the food manufacturing industries need to live up to the requirements of the consumers by producing good quality products with no toxic composition or adulteration. The manufacturers also required a few stepwise food processing structures/systems to maintain that quality. In addition to that, the industries need to follow a few food evaluation procedures to ensure that the products are suitable for consumption. There are few methods by which the processing can be performed those are: liquid chromatography (HPLC), liquid chromatography mass spectrometry (LCMS), enzyme linked immunoassay (ELISA) and polymerase chain reaction (PCR) [1, 2]. Basically, the food we consume when adulterated can cause a severe damage to our body. It is to be mentioned that there are few bio-sensing technologies which are considered to be cost effective, rapid and involve label-free techniques. Furthermore, bio-sensing technologies can be tailored to build suitable and compact device for segregation of unhealthy components from food products by lowering the detection limit. Amongst biosensors, optical biosensors are most commonly used biosensors. Optical biosensors are basically used in optical field interaction techniques to detect the bio-recognized elements. Surface plasmon resonance biosensors and localized surface plasmon resonance (LSPR) sensors are considered to be the most commonly used methods for their label free techniques.

2 Details of Surface Plasmon Resonance and Localized Surface Plasmon Resonance

Surface plasmon resonance and localized surface plasmon resonance based sensors are widely used for their handy and cost effective techniques. Moreover, these two sensing techniques are considered to be more selective, sensitive and stable towards the change in adjacent media near to the substrate. The technologies involved improving the quality of food utilizing SPR and LSPR is going to be discussed in details in this chapter.

2.1 Details of Surface Plasmon Resonance Sensors

SPR biosensor is made up of two key components. One is the optical system for excitation and another one is the interrogation of surface plasmons. In SPR biosensors to identify the bio-molecular recognition element and capture target analyte present in sample, biomolecular recognition elements are immobilized onto the surface of a metal layer. Surface Plasmon Resonance (SPR) is primarily employed for detecting biomolecules, food allergens, adulterants, microbes, toxins, antibiotic residues, pesticide and herbicide residues, and genetically modified foods. A number of researchers have focused on food analysis based in SPR technique and is featured as one of the popular techniques or even as the primary method for studying allergens, mycotoxins and antibiotics. The working principle of surface plasmon resonance is based on Kretschmann configuration. In the Kretschmann configuration (also known as the Kretschmann configuration), a thin metal film (mainly *gold*, *silver*, or *copper*) is deposited onto the hypotenuse face of a total-reflection prism (Fig. 1). The incident light passes through the substrate (viz. prism or glass) as well as the metal film that leads to excite surface plasmons on the opposite side of the film. This setup allows the researcher to study surface plasmon resonance (SPR). SPR occurs only when electrons in the thin metal film become excited by light directed at a specific angle of incidence. These excited electrons create a surface plasmon polariton, a non-radiative electromagnetic surface wave, that propagates parallel to the metal–dielectric interface. The plasmons are highly sensitive to changes in this boundary, such as the adsorption of molecules on the metal surface. The working principle of this technique can be segmented in three parts those are: (a) *Angle of Incidence*: The incident light strikes the metal film at a specific angle (known as the *SPR angle*). This angle depends on the refractive index of the surrounding medium. (b) *Resonance*: When the angle of incidence matches the condition for SPR, surface plasmons are excited. At this resonance angle, the reflected light intensity decreases significantly (resulting in a dip in the reflected light spectrum). (c) *Detection*: Researchers can measure this dip in reflected light intensity to detect specific substances (analytes) that interact with the metal surface. SPR biosensors have been developed for detecting biomarkers and other important molecules as well.

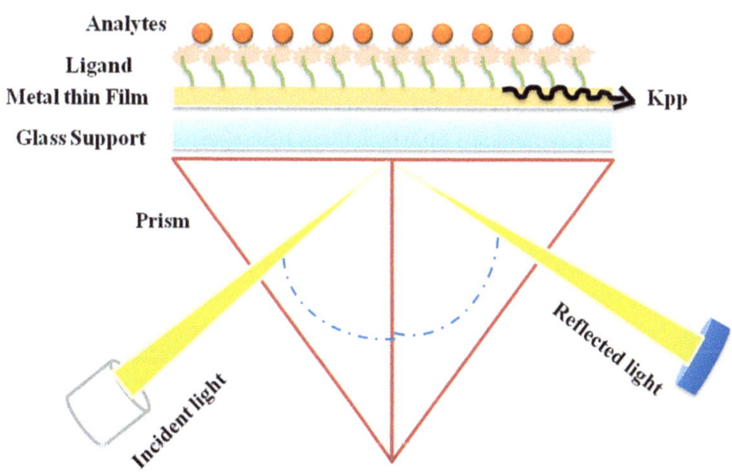

Fig. 1 Kretschmann configuration used for SPR sensors

2.1.1 Working Principle of SPR Sensor

Surface plasmons are considered to be formed by excitation of the free conduction electrons on a metal surface by p-polarized light. The SPs are formed at the interface of a metal and the respective dielectric material. The plasmons are the harmonic oscillations of the excited electrons at the metal interface. The interaction between the p-polarized electromagnetic radiation tend to produce surface plasmons only when the free electrons interact with the excited free electrons and lead to a progressive wave at the metal–dielectric interface. The progressive wave is also considered as electromagnetic wave. This is also considered to be the interaction of the evanescent wave field at the interface between the metal-dielectric and the surface plasmons. The photons of electromagnetic wave can excite the plasmons only when the resonant condition occurs i.e., when the frequency of the SPs and photons are equal. This phenomenon is called the surface plasmon resonance and the angle of incident at resonance is called resonant angle or SPR angle [1–4]. The complete phenomenon change due to change in refractive index (RI) adjacent to the evanescent wave field formed at the interface. Along with the above mentioned conditions it also important to note that the resonant angle dependents also on the wavelength or frequency of the incident light and the refractive index of the dielectric medium. Further, SPR curve also depends upon the thickness of the metal coating onto the sensing substrate as well. As the surface at the interface of a SPR sensor respond to the change in refractive index of the adjacent medium thus the surface is also responsive towards biomolecular interactions. The metal coating that are basically used during SPR phenomenon are: silver, gold, copper and aluminum out of these metals silver and gold are considered to be more sensitive and stable. A sensor build with the help of SPR phenomenon can also be used to detect the changes as these are responsive towards the aforementioned factors. In addition to that, change in temperatures can also have an impact on refractive index of the adjacent media as well. SPR sensors can be utilized to study the biomolecular interaction, food items and biomedicine as well which are responsive towards the change.

There are few more parameters on which the sensing performance of SPR sensors depends those are: wavelength of the incident light, resonant angle, incident angle, intensity of light, phase shift and phase/angular modulation. The parameters affect on to the surface plasmon wave and lead to alter the propagation constant. The change in propagation constant can also be recorded to observe the variations. There can be few different types of SPR sensors based on the phase, angular and intensity modulation. As the SPR sensors depend upon angular variations the change in phase and intensity of the source are the key parameters that lead to provide change in the output. SPR relies on total internal reflection (TIR), a phenomenon occurring when light passes from a denser medium (e.g., glass or prism) to a less dense medium (e.g., air or liquid). When light hits the interface at an angle greater than the critical angle, it reflects back into the denser medium instead of refracting out. Also, at the resonant angle, the incident light energy couples with surface plasmons (collective oscillations of electrons) at the metal–dielectric interface. This coupling leads to a reduction in reflected light intensity, creating a dip in the reflectance spectrum. The position of this dip is sensitive to changes in the refractive index near the sensor surface [1–4].

In most of the SPR based sensing system the Kretschmann configuration is used in which metal film viz. gold and silver is coated at the interface of the two dielectric media. This sensing system follows total internal reflection where the light needs to travel from higher refractive index (n_1) to lower (n_2) refractive index. At the medium of lower refractive index (n_2) side the evanescent waves generate under the condition of total internal reflection. The metal coating at the interface with suitable thickness at the interface can lead to enhance the evanescent wave penetrating to the gold film and exists in the second medium (n_2). The magnitude of the propagating wave vector can be expressed as follows:

$$k = \frac{2\pi}{\lambda} n_1 \sin\theta \tag{1}$$

where, λ is considered to be the wavelength incident light, n_1 refractive index of the first medium and θ is the incident angle.

The propagation of surface electromagnetic wave whose propagation is basically confined at the metal–dielectric interface depends upon the surface plasmons. As the surface plasmons are considered as quanta of plasma thus the magnitude of the wave vector of surface plasmons depends upon the second medium and the gold film (i.e. n_2 and n_g). The magnitude of the propagating wave vector of surface plasmon can be expressed as follows [3, 4]:

$$k_{SP} = \frac{2\pi}{\lambda} \frac{n_2^2 n_g^2}{n_2^2 + n_g^2} \tag{2}$$

where, n_2 is the refractive index of second medium and n_g is the refractive index of the gold film.

As mentioned earlier the surface plasmons can be excited by the evanescent wave at the metal dielectric interface which is surface plasmon resonance phenomenon that leads to decrease the intensity of the reflected light sharply. The excited surface plasmon includes energy conversion to photons due to the decays intensity. For SPR phenomenon the propagation vector $\mathbf{k_{sp}}$ must be equal to $\mathbf{k_{evanascent}}$. The angle required for resonance can be expressed as follows using Eqs. 1 and 2:

$$\theta_{SPR} = \mathrm{Sin}^{-1}\left(\frac{1}{n_1}\sqrt{\frac{n_2^2 n_g^2}{n_2^2 + n_g^2}}\right) \tag{3}$$

θ_{SPR}, is considered here to be the angle required for the resonance is related to n_2 (if n_1 and n_g are constant). Therefore, the monitoring of the θ_{SPR} change can be used to analyze the adsorption—desorption activities that take place on the gold surface.

2.2 *Details of Localized Surface Plasmon Resonance Sensors (LSPR)*

LSPR sensors are a type of optical sensor that detects changes in the refractive index near a metal surface due to the interaction of light with surface plasmons. These sensors offer high sensitivity, label-free detection, and real-time monitoring of biomolecular interactions. The working principle of the localized surface plasmon resonance is when a metal nanostructure (such as *gold* or *silver* nanoparticles) is irradiated with light the electromagnetic field of the light induces a collective coherent oscillation of the nanostructure's surface conduction electrons. The electric field associated with surface plasmons is confined near the nanoparticle surfaces and decaying exponentially over a few nanometers. This oscillation is in resonance with the specific frequency of the incident light. If the frequency of the incident light matches the natural frequency of the surface plasmons then only the resonance condition occurs. The detection mechanism of the LSPR sensor is it measures changes in the adjacent refractive index near the metal surface. When analyte molecules (e.g., proteins, DNA) bind to the sensor surface, they alter the local refractive index. The change in RI influences the resonance conditions of the surface plasmon waves. By monitoring the shift in the resonance angle or wavelength, we can detect binding events. There are few applications for which LSPR techniques are chosen for biosensing applications. The studies that can be done using LSPR are (a) *Biomolecular Interaction Studies*: Quantifying binding kinetics and affinities, (b) *Chemical Sensing*: Detecting minute changes, including rapid chemical reactions, (c) *Biological Monitoring*: Studying molecular activities and cellular processes. Moreover, this technique has high sensitivity, label-free and real-time monitoring i.e., LSPR sensors can detect small changes in refractive index. This technique does not require fluorescent or radioactive labels and it provides dynamic information during interactions. LSPR is also highly sensitive towards the size, shape and composition of the nanoparticles as well. Thus by altering these parameters a suitable sensors can be built for sensing purposes.

2.2.1 Working Principle of LSPR Sensing Technique

Localized surface plasmon resonance tends to alter the output with interacting with surrounding medium. Also, the peak wavelength generate is direct dependence with the surrounding media which can be given by the Drude model [5–7].

$$\varepsilon_r = 1 - \frac{\omega_p^2}{\omega^2 + \gamma^2} \tag{4}$$

where, ε_r is the real part of dielectric function (ε) of the plasmonic material, ω_p is the plasma frequency, ω denotes the angular frequency of radiation and γ is the damping factor of bulk metal. For, visible and infra-red regions $\gamma << \omega_p$ thus the above equation becomes,

$$\varepsilon_r = 1 - \frac{\omega_p^2}{\omega^2} \tag{5}$$

where, α is the polarizability of small particles whose size is much comparable to the wavelength of light can be given by,

$$\alpha = 3\varepsilon_0 V \frac{\varepsilon - \varepsilon_s}{\varepsilon + 2\varepsilon_s} \tag{6}$$

where, ε_0 denotes permittivity of free space, V is the volume of nanoparticles and ε_s is the dielectric constant of surrounding media. Resonance occurs when polarizability attains maximum value thus the resonance condition is $\varepsilon = -2\varepsilon_s$.

Now using Eq. (5) and the resonance condition, we can find out the frequency of the LSPR peak (ω_{max}) and given below:

$$\omega_{\max} = \frac{\omega_p}{(1+2\varepsilon_s)^2} \tag{7}$$

Taking $\lambda = 2\pi\varepsilon/\omega$ and $\varepsilon_s = n_s^2$, Eq. (7) becomes,

$$\lambda_{\max} = \lambda_p \sqrt{2n_s^2 + 1} \tag{8}$$

where, λ_{max} represents LSPR peak wavelength and λ_p is wavelength corresponding to plasma frequency of the metal, expression (8) shows the linear relationship between LSPR peak wavelength and RI of the surrounding medium.

The sensitivity (S) of an LSPR sensor is defined as a change in peak wavelength maximum per unit change in surrounding RI and can be calculated as,

$$s = \frac{\delta\lambda}{\delta n} \tag{9}$$

It is to be noted that sensing phenomenon depends upon few key factors those are the material composition, shape, size, and density of the nanoparticles. Thus, LSPR wavelength changes according to the mentioned parameters as well that lead to alter the RI index of the surrounding medium. Thus fiber coated with nanoparticles can also be utilized to study few applications such as the bio-sensing application, medicine or clinical diagnosis, and fuel quality detection in industries [7–10] (Fig. 2).

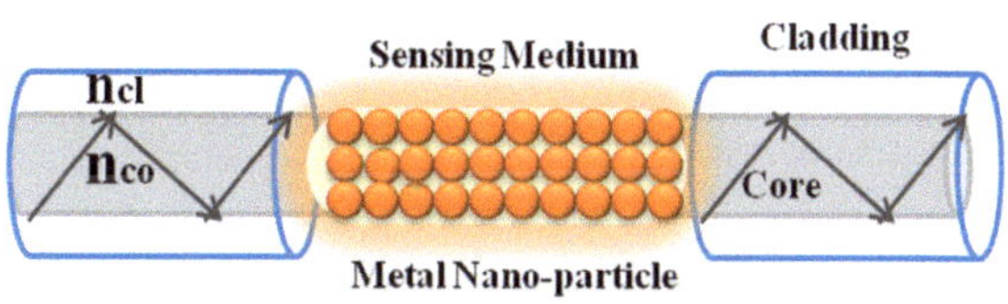

Fig. 2 Optical fiber coated LSPR sensors

2.3 Food Adulteration Detection Using SPR Techniques

In SPR technique detection of biological sample for any adulteration or added preservative used a well known Kretschmann type prism coupling technique along with change in resonance angle with fixed incident wavelength. Basically, the SPR optical system comprises of an excitation light source, prism and a metal film (~50 nm thick) coated onto the surface of the prism proposed by Zhang et al. (2016). According to them the evanescent wave traverses the metal film, causing the SPR phenomenon at the interface between the metal film and food allergens captured on its surface. Also, it is to be noted that SPR biosensors enumerates food allergens by monitoring the adsorption and dissociation of their binding to receptors in real-time and recording the SPR signal value by analyzing the dissociation rate constant. Additionally, SPR biosensors can be regenerated by disrupting the analyte receptor binding with a regeneration solution (high-salt or acidic) was proposed by Wammes et al. (2015). The excitation of an SPR depends on a number of parameters that includs the incident angle, incident wavelength, metal film, prism refractive index and medium refractive index. Based on different detection parameters, SPR biosensors can be categorized into four methodologies for measuring the SPR signal such as angular modulation (by Bevenot et al. 2001; Slavik et al. 1998), wavelength modulation (by Manuel et al. 1993), intensity modulation (by Nelson et al. 1996) and phase modulation (by Esteban et al. 2000; Kazuyoshi and Koji 2002). The angular modulation in SPR biosensor measures the angle shift at resonance under a fixed incident wavelength. The wavelength modulation in SPR biosensor monitors the wavelength shift at resonance under a fixed angle of incidence. The intensity modulation in SPR biosensor detects changes in reflected light intensity to obtain the refractive index-reflected light intensity curve at a fixed incident wavelength and angle. The phase modulation in SPR biosensor detects the phase change between the reflected and incident light. In addition to that, in SPR technique the sensors are basically worked on the principle of binding onto the surface of adulteration or added preservatives present on food samples. The toxins can also be detected by targeting specific gene sites of proposed samples as well. Because of their level free techniques, SPR has been used in many samples for detection of food preservatives such as in honey, milk and egg etc. To detect the added preservative on honey sample by Vikas et al. have used silver coated SPR prove to improve glucose and fructose sensitivity. Further, Oh et al. (2019) have used and improved LSPR technique to detect melamine in milk powder [11–20] (Tables 1 and 2).

Table 1 Methods used for food detection

Methods/device	Uses	Components used
Fiber optic SPR (FOSPR)	Real time detection Remote sensing	• Light source • Sensing probes: sensitive layer, metal film and core • Sample cell • Detector
SPRI	Real time detection High throughout analysis	• Light source • Kretschmann-type prism • Metal film • Sample cell • Charge coupled device (CCD) imaging
TSPR	High sensitivity Real time detection	• Light source • Grating Substrate • Sample cell • Metal film • Detector
LSPR	Device miniaturization Multiplexed detection	• Light source • Noble metal nanostructures • Sample cell • Detector

3 Application of SPR and LSPR Towards Food Detection Technique

3.1 Application Utilizing SPR Sensor

When a biomolecule (e.g., protein, DNA) binds to the sensor surface, it alters the local refractive index. This change affects the SPR condition, shifting the resonant angle. By monitoring the reflected light intensity, we can detect binding events in real time. The stronger the binding, the larger the shift in the SPR signal. Further, SPR phenomenon occurs at particular angle can be also be altered by introducing different analytes adjacent to the metal film, thus can be utilized to sense the change. The change in chemical compositions and biochemical compositions nearby the metal surface within the range of plasmonic field can lead to change in the resonance with the plasmons. As discussed earlier surface plasmon resonance can also be termed as affinity senor based on their selectivity and affinity towards binding of specific molecule. Also, bio-recognition element is immobilized onto the metal surface at which surface plasmons exists. In presence of suitable analyte mixture binding can occur at the interface which lead to increase the refractive index at the contact surface and can be recognized by the SPR sensor. Based on the phenomenon of SPR there is a specific technique i.e., MIP has been first coined out by few researchers. Molecularly Imprinted Polymers (MIPs) are synthetic materials designed to selectively recognize

Table 2 Detection of different food adulterations using SPR and LSPR sensors

Biosensors	Sample	Capturing ligands	Senor characteristics
SPR	1. Milk	1. NanoMIPs	1. LOD—127 ± 97.6 ng/ml (Ashley et al. 2018)
	2. Milk	2. Antibody	2. LOD—0.164 μg/ml (Ashley et al. 2018) LOD—0–1000 ng/ml (Billakanti et al. 2010) LOD—2.8 × 10–7 M 5.0 × 10–8 M (Tomassetti et al. 2013 LOD—57.8 ng/ml (Ashley et al. 2017)
	3. Egg	3. Antibody	3. LOD—0.03–0.2 μg/ml (Pilolli et al. 2015) LOD—0.2 μg/ml (Pilolli and Monaci 2016) LOD—0.035 μg/ml (Nylander et al. 1983)
	4. Peanut	4. Antibody	4. LOD—0.77 ng/ml (Wu et al. 2016) LOD—Mohammed et al. 2001)
	5. Fish	5. Antibody	5. LOD—0.11 mg/kg LOD—0.39 mg/kg
LSPR	1. Chili powder, egg, sauce	1. Sudan I	1. LOD—0.001 μM LOD—0.01 μM/L 84–110 [11, 11–20]
	2. Biological samples	2. Dopamine	2. LOD—0.011 μM [11]
	3. Orange sweetie/green sweetie/cocktail margarita	3.Tartrazine	3. LOD—2 nM [11]
	4. Milk	4. Melamine	4. LOD—3 × 10–13 mol/L [11, 11–20]
	–	5. Paraben	5. LOD—0.706 μM

and bind specific target molecules (analytes) based on their shape, size, and functional groups. The process involves creating a polymer matrix with binding sites that mimic the target molecule's structure. MIPs are widely used in various applications, including sensors, separation, and drug delivery. MIP-Based SPR sensors are widely used to create highly selective and sensitive detection platforms. A thin gold film is coated with an MIP layer contains imprinted binding sites specific to the target analyte. When the analyte binds to the MIP leads to a change in refractive index at the sensor surface thus alter the SPR signal [12–16].

3.2 Application Utilizing LSPR Sensor

Detection of food contamination and pesticides are essential for health benefits. Moreover, presence of harmful mixtures and pesticides can add loss to food industries as well. To assess the food quality with the use of nano-materials based sensors are considered handy and rapid.

There are few optical sensors are considered to be suitable for detection methods of food ingredients in minute quantities. As optical sensors using nano-materials are rapid, accurate and handy in detecting the minute adhere in food. These are also considered to be more accurate and quick in analyzing harmful chemicals in food ingredients than traditional bulky methods such as HPLC, GC–MS etc. Pesticides are one of the major concerns for the framers and the food growing industries now a day's thus to detect those adhere we need a rapid technique. This is the reason behind using nano-materials and LSPR techniques. There are a number of optical methods for detection out of which fluorescence quenching method is commonly used. In this technique pesticides can be detected by direct interaction with quantum dots. Moreover, different composite quantum dots such as QDs-QDs [11–15] and QDs-AgNPs can solely detect multi-component with high selectivity and sensitivity. Ratiometric fluorescence is also one of the techniques to detect pesticides with more sensitivity. Nanomaterials in sensors are also sometimes combined with enzymes which can catalyze certain reactions that lead to improve the specificity and selectivity of certain biosensors to detect few pesticides. Those composite materials in sensors are considered to be suitable for detection of pesticides based on organophosphorus (OPP). Combination of nanoparticles, aptamer and molecular imprinting polymers (MIP) are considered more specific towards detection of pesticides as well. Also, MIP modified quantum dots were used for determination of thiabendazole (TBZ) as fluorophores in optical sensors. MIPs can also be utilized for detection of 2, 4-D based on ratiometric fluorescence using QDs. Noble metal nanoparticles such as AgNPs and AuNPs are used in sensors for detection of tetracyclines (TCs) that present in antibiotics. Sensors array can also be used for more detailed analysis of antibiotics present in food [17–20].

4 Importance of Analysis of Food Products

Detecting melamine in food products is crucial for ensuring food safety. Here are some methods used for melamine detection: Liquid Chromatography Triple Quadrupole Tandem Mass Spectrometry (LC-MS/MS): This method involves an initial extraction with 2.5% aqueous formic acid, followed by filtration, centrifugation, and dilution steps. Melamine is detected in positive ion mode, and cyanuric acid (often associated with melamine) is detected in negative ion mode. Analyte concentrations are calculated using external standard calibration. Fluorescence-Based Chemical Sensors: These sensors offer a simple and fast way to detect melamine in food samples. However, they may not be as precise as LC-MS/MS. Other Techniques: Chemiluminescence, immunoassays, and various optical spectroscopic techniques have also been explored for melamine detection. Researchers continue to innovate in this area to improve detection accuracy and efficiency. Remember that melamine adulteration can have serious health consequences, so reliable detection methods are essential to maintain food authenticity and safety. Non-targeted detection methods play a crucial role in identifying unknown adulterants or foreign components in food products. Let's delve into the specific example of detecting adulterated vegetable cooking oils using a rapid optical sensor based on fluorescence quenching of CdSe/ZnS quantum dots (QDs): Fluorescence Quenching Sensor: This method utilizes CdSe/ZnS QDs, which are semiconductor nano-crystals with unique fluorescence properties. When these QDs come into contact with certain substances (such as adulterants), their fluorescence is quenched. The sensor detects the presence of refined oil (an adulterant) in soybean oil by measuring the change in fluorescence within a short time (2 min). Importantly, this technique doesn't require any complex sample preparation, making it rapid and efficient [18–20].

Food can be analyzed using few techniques such as gas or liquid chromatography and spectroscopic techniques. Chromatographic techniques offer better detection performance, high separation efficiency but are considered to be costlier. Spectroscopic techniques are Raman, UV–Vis, NIT are quick to operate but shows poor detection specificity. Owing to the aforementioned challenges these are not popularly used in food safety analysis. Like SPR sensor LSPR sensors also have the ability to detect composition added to food. Thus, nano sensors are also widely used for their rapid detection, handy techniques, low cost and high sensitive towards the alterations of adjacent environment. Since, now a day's food safety is vital for human being as well as for industrial growth. Thus maintaining the safety and the quality is the key factor to observe during cultivation and industrial manufacturing. In addition to that we need to look after the addition of preservative and contamination as well while going for quality check. There can be two different aspects related to food safety such as exogenous and endogenous. Endogenous food safety refers to as the presence of pesticides, few veterinary drug and heavy metals, whereas endogenous food safety refers to as the presence of biological toxins and the harmful chemicals.

5 Conclusion

Modern technologies such as nanofabrication and nano-encapsulation offers innovative solutions for creating healthy foods with bio-actives and achieving targeted controlled release in the gut. The technologies are considered to be novel as compared to any other modern techniques. These advancements in technologies have the capacity to enhance the nutritional value of food products and can also be utilized for advancement in detecting adulteration. Additionally, nanomaterials can also be directly incorporated into food matrices or food contact materials to improve food preservation. By incorporating nanomaterials into food products, we can develop nano-composites with properties such as barrier enhancement, antimicrobial effects, antioxidant activity and oxygen scavenging. Nanomaterials also support circular bio-economy strategies by utilizing agro-industrial waste and food processing by-products, aligning with sustainability goals and resource optimization. Continuous advancements in nano-science and technology are expected to facilitate their widespread adoption.

References

1. Y. Tang, X. Zeng, J. Liang, Surface plasmon resonance: an introduction to a surface spectroscopy technique. J. Chem. Educ. **87**(7), 742–746 (2010)
2. A.D.S. Arcas, F.D.S. Dutra, R.C. Allil, M.M. Werneck, Surface plasmon resonance and bending loss-based U-shaped plastic optical fiber biosensors. Sensors **18**(2), 648 (2018)
3. M.T.I. Ansari, S.K. Raghuwanshi, S. Kumar, Recent advancement in fiber-optic-based SPR biosensor for food adulteration detection—a review. IEEE Trans. Nanobiosci. **22**(4), 978–988 (2023)
4. H. Agrawal, A.M. Shrivastav, B.D. Gupta, Surface plasmon resonance based optical fiber sensor for atrazine detection using molecular imprinting technique. Sens. Actuators, B Chem. **227**, 204–211 (2016)
5. W. Wang, Y. You, S. Gunasekaran, LSPR-based colorimetric biosensing for food quality and safety. Compr. Rev. Food Sci. Food Saf. **20**(6), 5829–5855 (2021)
6. M. Chauhan, V.K. Singh, Review on recent experimental SPR/LSPR based fiber optic analyte sensors. Opt. Fiber Technol. **64**, 102580 (2021)
7. J.G. Ortega-Mendoza, A. Padilla-Vivanco, C. Toxqui-Quitl, P. Zaca-Morán, D. Villegas-Hernández, F. Chávez, Optical fiber sensor based on localized surface plasmon resonance using silver nanoparticles photodeposited on the optical fiber end. Sensors **14**(10), 18701–18710 (2014)
8. R. Dutta, R. Bharadwaj, S. Mukherji, T. Kundu, Study of localized surface-plasmon-resonance-based optical fiber sensor. Appl. Opt. **50**(25), E138–E144 (2011)
9. C. Caucheteur, T. Guo, J. Albert, Review of plasmonic fiber optic biochemical sensors: improving the limit of detection. Anal. Bioanal. Chem. **407**, 3883–3897 (2015). https://doi.org/10.1007/s00216-014-8411-6
10. H. Luo, Y. Huang, K. Lai, B.A. Rasco, Y. Fan, Surface-enhanced Raman spectroscopy coupled with gold nanoparticles for rapid detection of phosmet and thiabendazole residues in apples. Food Control **68**, 229–235 (2016)
11. J. Zhou, Q. Qi, C. Wang, Y. Qian, G. Liu, Y. Wang, L. Fu, Surface plasmon resonance (SPR) biosensors for food allergen detection in food matrices. Biosens. Bioelectron. **142**, 111449 (2019)
12. Y.N. Wu, P. Liu, J.S. Chen, Food safety risk assessment in China: Past, present and future. Food Control **90**, 212–221 (2018)
13. A. Liu, L. Shen, Y. Tan, Z. Zeng, Y. Liu, C. Li, Food integrity in China: insights from the national food spot check data in 2016. Food Control **84**, 403–407 (2018)
14. R.E. Hunter Jr., A.M. Riederer, P.B. Ryan, Method for the determination of organophosphorus and pyrethroid pesticides in food via gas chromatography with electron-capture detection. J. Agric. Food Chem. **58**(3), 1396–1402 (2010)
15. N. Yoshioka, K. Ichihashi, Determination of 40 synthetic food colors in drinks and candies by high-performance liquid chromatography using a short column with photodiode array detection. Talanta **74**(5), 1408–1413 (2008)

16. H.Y. Fu, H.D. Li, B. Wang, J.L. Cai, J.W. Guo, H.P. Cui, Y.J. Yu, Quantification of acid metabolites in complex plant samples by using second-order calibration coupled with GC-mass spectrometry detection to resolve the influence of seriously overlapped chromatographic peaks. Anal. Methods **8**(4), 747–755 (2016)
17. Z. Mohammadi, S.M. Jafari, Detection of food spoilage and adulteration by novel nanomaterial-based sensors. Adv. Coll. Interface. Sci. **286**, 102297 (2020)
18. J.M. Soon, A.K. Brazier, C.A. Wallace, Determining common contributory factors in food safety incidents—a review of global outbreaks and recalls 2008–2018. Trends Food Sci. Technol. **97**, 76–87 (2020)
19. L. Manning, P.A. Luning, C.A. Wallace, The evolution and cultural framing of food safety management systems—where from and where next? Compr. Rev. Food Sci. Food Saf. **18**(6), 1770–1792 (2019)
20. P. Zhou, O. Hu, H. Fu, L. Ouyang, X. Gong, P. Meng, Y. Wang, UPLC–Q-TOF/MS-based untargeted metabolomics coupled with chemometrics approach for Tieguanyin tea with seasonal and year variations. Food Chem. **283**, 73–82 (2019)

Optical Techniques/Probes for the Detection of Antibiotic Residues in Milk and Dairy Products

Shalini Singh, Mehul Chudasama, Shreya Singh, Falguni Patra, and Raj Kumar Duary

Abstract Optical techniques/probes are increasingly being used for the analysis and quality control of milk and products produced from milk. Optical techniques of antibiotic detection have been extensively used during the last twenty years due to their rapid exposure cycles for detection, high sensitivity, and simple procedure, which are mainly researched for water and food quality analysis. These optical detection methods are advantageous over conventional methods for the analysis of antibiotic residues in milk and milk products. Antibiotic residues are known as the remnants of antibiotics that remain in food commodities such as milk and dairy products after they have been used to treat cattle for various diseases or defects and to promote the growth of livestock. Major residues present in milk reported beta-lactam, tetracyclines, fluoroquinolones, sulphonamides and aminoglycosides. The presence of such residues poses significant health risks, making their detection and quantification paramount for ensuring food safety. Present optical analytic methods are rapid and have ease of detection. These advancements are important to maintain food safety by controlling the accumulation and transfer of antibiotics in the food chain. Considering the potential hazards of antibiotic residues and the necessity for optical detection methods, and possible advancements and expansions in the detection of antibiotic residues, this current chapter has been proposed.

Keywords Antibiotic residues · Optical techniques · SPR · SERS · Biosensors · Detection methods

S. Singh · S. Singh · R. K. Duary (✉)
Department of Dairy Science and Food Technology, Institute of Agricultural Sciences, Banaras Hindu University, Varanasi, India
e-mail: rkduary@bhu.ac.in

M. Chudasama
Department of Food Technology, Parul Institute of Technology, Parul University, Vadodara, Gujarat, India

F. Patra
Mansinhbhai Institute of Dairy and Food Technology, Dudhsagar Dairy Campus, Mehsana, Gujarat, India

R. Biswas (ed.), *Optical Techniques for Assessing Food Adulterants*, Progress in Optical Science and Photonics 35, https://doi.org/10.1007/978-981-96-8142-6_5

1 Introduction

1.1 Importance of Milk and Dairy Products in the Human Diet

Milk and dairy products play a critical role in the human diet due or their rich nutrient content, providing minerals and essential vitamins and protein necessary for overall health. Milk is the primary source of dietary calcium which is required for various functions like building and maintaining strong bones and teeth. Sufficient calcium intake during childhood, adolescence, and early adulthood can help to increase bone mass and reduce the risk of osteoporosis during maturity [1]. Other than calcium, protein and vitamin D, milk also contains other essential nutrients such as vitamin A, vitamin B_{12}, riboflavin, potassium and phosphorus which are crucial for many bodily functions. Milk and milk products are essential for the growth and development of children and adolescents. These nutrients are necessary for bone growth, cognitive development and overall health [2].

1.2 Overview of Antibiotic Use in Dairy Farming

Antibiotics are chemical substances derived from microorganisms, such as molds or bacteria, that can inhibit the growth of or eliminate other microorganisms, typically bacteria, and treat infections. Antibiotics may be characterised under two types: broad-spectrum antibiotics which are effective against a wide range of infections, and others, called narrow-spectrum antibiotics, target only a limited number of bacterial types. Antibiotics are mostly used for animals in dairy farms for the prevention of disease and to treat cattle. Common diseases in cows are mastitis (udder inflammation), metritis (uterine infection), and respiratory infections, which may be treated with antibiotics to ensure the health and welfare of the animals [3]. Tiseo and co-workers studied the use of antibiotics in animal food from the year 2017 to 2030, results presented that the sales are expected to rise by atleast 11.5% by 2030 [4]. These are also used at subtherapeutic levels to promote development and improve feed effectiveness in cattle. This creates problems about the development of antibiotic-resistant bacteria. Prophylactic use of antibiotics: Dry cow therapy and post-surgical risk management both involve antibiotics for treatment [5].

1.3 *Residues, Antibiotic Residues and Maximum Residue Limit (MRL)*

A residue refers to any compound found in edible tissues as a result of drug use, including the drug itself, its metabolites, and any other substances produced in or on food due to the drug's application [6]. Generally, drugs that are administered in the body of an animal are excreted and removed from the system. Every metabolite is eliminated through urine and to a lesser extent through feces [7]. Even after the excretion these might persist in various food materials such as eggs, milk and meat. The U.S. Food and Drug Administration (FDA) defines residues as "the amount of drug or its metabolites remaining in the target tissue or edible product of a food-producing animal at the moment of slaughter or at the time the food is removed from the animal for human utilization." The U.S. Food and Drug Administration (FDA) outlines antibiotic residues as "the remnants or remains of antibiotic drugs or their metabolites that persist in the tissues or products derived from food-producing animals after the administration of antibiotics." The maximum residue limit (MRL) refers to the highest level or concentration of a drug or chemical considered safe and allowed by regulatory authorities in or on food or feed intended for animal or human consumption at a specified time. The FDA sets tolerance levels or maximum residue limits (MRLs) for specific antibiotics in various food products to ensure that residues do not endanger people's health.

2 Antibiotic Residues in Milk and Dairy Products

2.1 *Common Antibiotics Used in Animals Under Dairy Farming*

Antibiotics are low to medium (100–1500) molecular weight compounds. These can be classified in various ways, including by their source of origin, chemical structure, bacterial spectrum (narrow or broad), route of administration (oral, topical, injectable), or their mode of action (bactericidal or bacteriostatic). Table 1 describes the commonly used antibiotics, their effects in human body and chemical structure. Six main classes of antibiotics are commonly used for dairy farm animals, including tetracyclines, sulfonamides, penicillins, amphenicols, aminoglycosides, beta-lactams, and fluoroquinolones. These antibiotics are primarily administered to dairy cattle for various purposes, such as disease treatment, prevention, and prophylaxis, including use as growth promoters [8]. A list of common antibiotics used in milk and its products is depicted in Table 2 along with their concentration detected in the given product.

Table 1 Antibiotics in dairy farms: structure, effects in human beings, maximum residue limit

Antibiotics	Structure	Effects in human	MRL*	References
Oxytetracycline	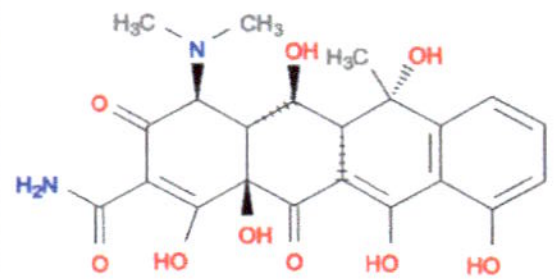	Inflammatory, teeth staining, repression of bone growth, hepatoxicity	100 μg/L	[9, 10]
Amoxicillin	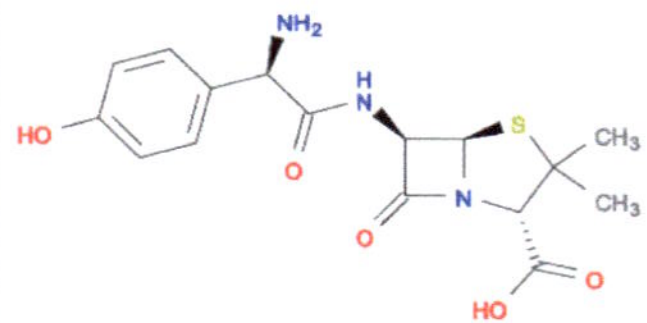		4 μg/kg	
Tetracycline	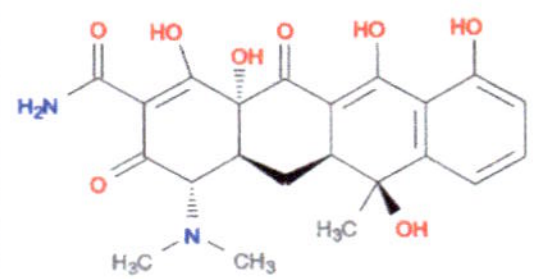	Liver damage, tooth yellowing, hepatotoxicity, nephrotoxicity	100 μg/L	[11, 12]
Fluoroquinolones	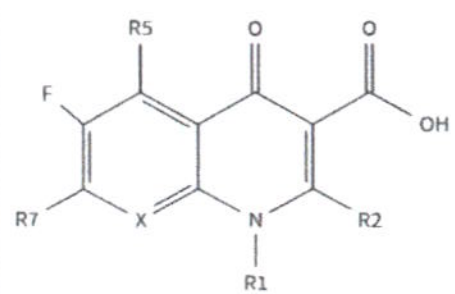	Neurotoxic, retinal degeneration,		[13]

(continued)

Table 1 (continued)

Antibiotics	Structure	Effects in human	MRL*	References
Gentamicin	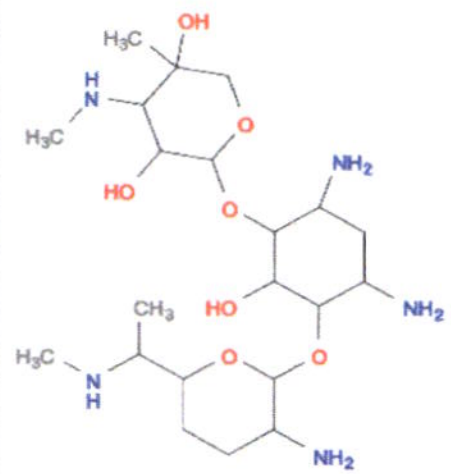	Nephrotoxicity, Ototoxicity	200 µg/L	[14]
Enrofloxacin	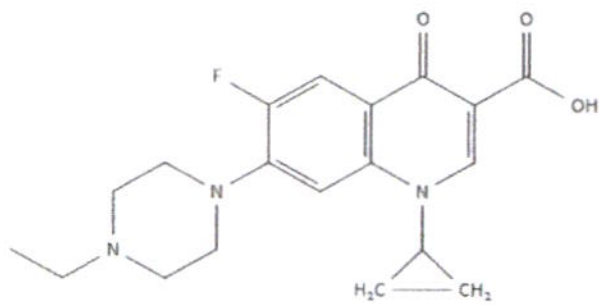	Hypersensitivity reactions	100 µg/kg	WHO & FAO (2023)
Penicillin G	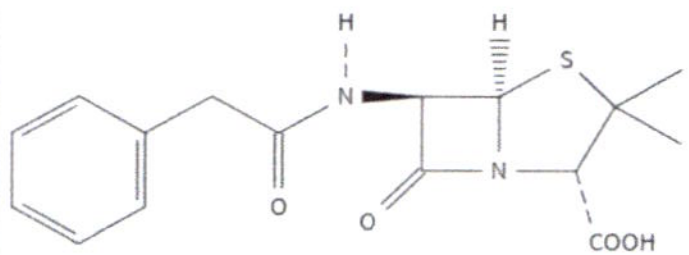	Hypersensitivity reactions, allergic reactions to skin, gastrointestinal problems	4 ng/mL	[15]

(continued)

Table 1 (continued)

Antibiotics	Structure	Effects in human	MRL*	References
Quinolones	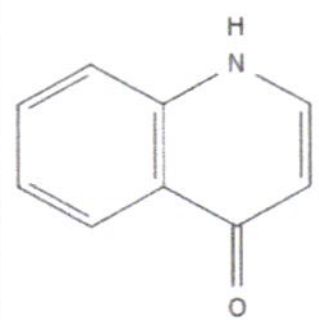	Tendon rupture, Tendinitis, severe hepatic toxicity	30–100 μg/kg	[16]
Streptomycin	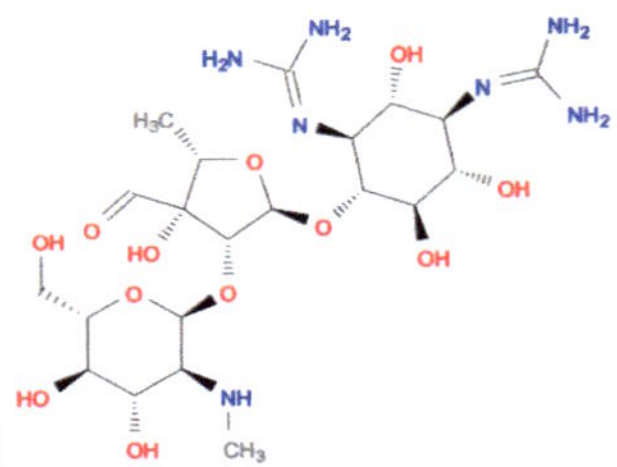		200 μg/kg	[17]

* MRLs values are taken from Codex Alimentarius International Food Standards, FAO-WHO (2023)

Table 2 Common antibiotics used in dairy farms, and its concentrations using different methods of analysis for milk and milk products researched in various regions globally

Source	Antibiotics	Concentration	Method	Region	References
Pasteurized cows' milk	Oxytetracycline Ciprofloxacin Enrofloxacin Levofloxacin	6–43 μg/L Absent 14–31 μg/L 3–5 μg/L	HPLC–UV	Dhaka, Bangladesh	[18]
Raw cows' milk	Oxytetracycline Ciprofloxacin Enrofloxacin Levofloxacin	Absent 28–30 μg/L 20–25 μg/L 3–4 μg/L			
Raw Pasteurized	Oxytetracycline	150.6 ng/g 42.6	HPLC–DAD	Northwestern Himalayan, India	[19]
Raw Pasteurized	Amoxicillin	12.3 NIL			
Cheese	Enrofloxacin Ciprofloxacin	38 ± 2 μg/kg 135 ± 2 μg/kg	LC–MS-MS	Valencia, Spain	[20]
Pooled raw milk	Tetracycline Fluroquinolones Gentamicin	667.09 μg/kg 80.44 μg/kg 47.92 μg/kg	ELISA	Gujarat, India	[21]
Pooled raw milk	Oxytetracycline	201.00 ± 41.25 ng/mL 272.11 ± 53.21 ng/mL	ELISA HPLC	Kerala, India	[22]
Raw cow milk	Tetracyclines Enrofloxacin β-lactams Others/unknown antibiotics	1.82% 1.82% 2.42% 2.42%	Charm assay	Thrissur, India	[23]
Raw milk	Tetracycline Sulfadimidine Oxytetracycline Sulfamethoxazole	>0.1 μg/mL		Bihar, India	[24]
Fresh mik	Penicillin-G	~15.22 μg/L	HPLC	Oyo state, Nigeria	[25]
Fermented milk (nono)		~7.6 μg/L			
Cheese (wara)		~8.24 μg/L			
Milk	Amoxicillin	9.8–56.16 μg/mL	UHPLC	Chittagong, Bangladesh	[26]
Raw milk	Quinolones Sulphonamides	23.25 μg/kg 16.28 μg/kg	ELISA	China	[27]
Raw milk	Oxytetracycline Amoxicillin	116 μg/L 345 μg/L	TLC & UHPLC	Chattogram, Bangladesh	[28]

(continued)

Table 2 (continued)

Source	Antibiotics	Concentration	Method	Region	References
Mango milk		Absent		Patiya Upazila, Bangladesh	
Chocolate milk		Absent			
Strawberry milk		Absent			
UHT milk		Absent			
Pasteurized milk		Absent			

2.2 *How Antibiotic Residues End Up in Milk?*

Antibiotic residues get administered in milk products through contaminated milk. Intramammary infusions used to treat mastitis and during the drying-off period are the primary sources of contamination globally [29, 30]. Antibiotics are commonly administered to dairy cows to treat bacterial infections such as mastitis [31]. When antibiotics are given to cows, small amounts may be excreted in their milk, especially if the withdrawal period (time for the antibiotic to be eliminated from the cattle's system) has not been observed [32]. There are few other possible ways of contaminations:

(a) Antibiotics may contaminate milking equipment, surfaces or udders during the milking process. If proper sanitation and cleaning procedures are not followed, residual antibiotics can transfer to the milk [33].
(b) Tetracycline is used as food baths to treat food rot, which gets licked by cows afterwards.
(c) Milk from treated cow accidently incorporated into bulk milk
(d) Other types of intramuscular and intravenous treatments in cattle

Residues from antibiotic drugs exist in the milk and its products due to the limited detection facilities, improper monitoring system [34] and availability of fewer quick detection systems for antibiotic residue analysis.

2.3 *Risks Associated with Antibiotic Residues in Milk and Its Products*

Residues from antibiotic drugs persist in the milk and its products subjected to various rigorous processes such as thermal processing. Antibiotics are mostly utilised in dairy farming for the management and treatment of mastitis and other contagious diseases [35, 36]. Ingestion of milk that contains antibiotic residues can cause allergic reactions in individuals who are sensitive to these substances [37]. Low levels of antibiotic residues can trigger allergic responses in some people. More importantly low-level doses of antibiotic residues in food commodity consumed for longer period may result in development of toxic allergic reactions and carcinogenic effect [38]. Antibiotic residues in food product such as milk can contribute to the development of antibiotic-resistant bacteria in humans. Exposure to low levels of antibiotics through milk consumption can lead to resistant strains of bacteria in the gut, potentially leading to infections that are difficult to treat with common antibiotics [39]. Consumption of milk that contains antibiotic residues may disrupt the balance of beneficial bacteria in the gut, leading to gastrointestinal disturbances and potential long-term health effects [40]. When not used properly, without considering the medical history and are advised withdrawal period, antibiotics may be found as residues in the food matrix. Other than issues with human and animals, several studies have shown that antibiotic residues have a negative impact on the fermented milk industry. Some of the effects include longer rennet clotting times, the inhibition of lactic acid production, reduced diacetyl production [41, 42], delayed cheese ripening.

2.4 Antibiotic Resistance Among Bacteria and Its Health Implications

The primary factor contributing to the advent of antibiotic-resistant bacteria is the overuse of drugs in dairy farming and inadequate biosecurity measures on farms [43]. Common diseases occur within the dairy farm such as mastitis, foot diseases, respiratory issues, etc.). Mastitis is known as a bacterial disease that generally occurs within dairy farms among cattle. It results in huge economic loss. Unhygienic milking practices, inefficient treatment, lack of awareness, hindered detection of sub-clinical mastitis, etc. are some of the common contributing factors to the higher incidence of mastitis. Mastitis not only affects the quality of milk but is mostly concerns beyond the dairy farm. Especially, the use of antibiotics for its prevention which ends up as residues. Increasing consumer concern over antimicrobial residues, antibiotic resistance, milk quality, and animal welfare are some of the major reasons to address mastitis prevention and management (National Academy of Agricultural Sciences, New Delhi 2013).

2.5 Health Implications of Consuming Milk with Antibiotic Residues

Consuming milk with antibiotic residue can have several health implications for consumers, including allergic reactions, disruption of gut microbiota, and the development of antibiotic resistance. Humans are exposed to antibiotic residues through food when traces of drugs used to treat food animals remain in their muscle tissue or in animal-derived products like milk and meat [44]. Some of the major health implications related to antibiotic residues are mentioned below:

Allergic reaction: Antibiotics in milk can cause allergic reactions in sensitive individuals. Even low levels of antibiotic residues can trigger allergic responses in some people [45].

Gut microbiota disruption: consumption of milk with antibiotic residues can disrupt the balance of beneficial bacteria in the gut, leading to gastrointestinal disturbances and potential long-term health effects [40].

Development of antibiotic resistance: Exposure to low levels of antibiotics through milk consumption can select resistant strains of bacteria in the gut, potentially leading to infections that are difficult to treat with common antibiotics [46]. This is a significant concern as antibiotic resistance can lead to the ineffectiveness of antibiotics in treating bacterial infections.

Effects on children: Children may be more vulnerable to the health effects of consuming milk with antibiotic residues due to their developing immune and digestive systems [47].

Other Health Risks: Other potential health risks include the risk of teratogenic effects on the fetus, underdevelopment of teeth, bone marrow aplasia, chronic low-level exposure increasing the risk of cancer, and disturbances to the reproductive, immune, endocrine, and nervous systems [48].

3 Optical Techniques for Antibiotic Residue Detection

Optical techniques are the methods which use light (visible, infrared, ultraviolet, etc.) for several applications, such as imaging, sensing, analysis and some communications. These techniques are widely used across different fields namely biology, physics, engineering, etc. Optical techniques are non-destructive, non-invasive and require minimal sample preparation. These are considered highly sensitive having great precision and real-time monitoring [12]. It is used for non-contact measurement in various fields of research and study [49, 50]. Some optical analysis tools are compact, portable, and easy to be miniaturized to develop integrated devices. For analysis purposes, sample preparation is easy for techniques like electron microscopy or mass spectrometry. Optical fiber sensing methods for antibiotic detection integrates optical fiber substrates with optical immunoassay techniques and molecular spectroscopy detection methods. It also involves designing specialized fiber structures, such as tapering and film coatings, to enhance sensing performance. Fiber sensors equipped with SPR (Surface Plasma Resonance), fiber sensors integrated with LMR (Lossy Mode Resonance), tapered fiber sensors and phonic crystal type fiber sensors. Optical fiber-based sensors, as a novel sensing method, reduce volume and transmission loss while enhancing sensing performance through the design of specialized sensing structures.

3.1 *Surface Plasma Resonance Based Biosensors*

SPR-based biosensors are highly sensitive to molecular interactions which makes them ideal for the detection of the minute quantity of residues in milk and dairy products. The basic principle of this technique is to measure the variations in the refractive index which occurs near the sensor surface used for detection (Fig. 1). The surface sensor is produced by coating with a specific recognition element or material such as antigen, enzymes, aptamer, antibodies and molecularly imprinted polymer (MIP) films that selectively bind to the antibiotic molecule. SPR-based sensing technology is known for its high sensitivity, real-time monitoring and detection, multiple detection system and label-free sensing [51].

Principle of SPR: SPR occurs at the interface of two semi-infinite media and is defined as the collective plasma oscillation or charge density oscillation of free electron gas present in thin metallic films that is in resonance with applied monochromatic electromagnetic radiation using non-radiative excitation [52]. Surface plasmons (SP) are generated when p-polarized light stimulates the free conduction electrons on a

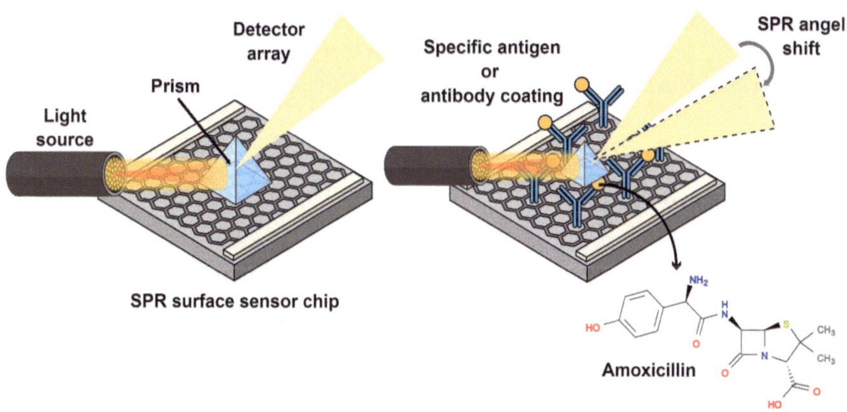

Fig. 1 Schematic diagram of the basic principle of surface plasma resonance type biosensor to detect antibiotic residues

metal surface. These plasmons occur at the interface between a metal and a dielectric material. The high-density electrons oscillate harmonically at this interface, giving rise to the term "plasmons." To efficiently produce SPs, a metal film is deposited on a prism and placed in contact with a dielectric material. When plane polarized electromagnetic waves interact with a metal–dielectric interface, they can generate surface plasmons. This occurs as the free electrons in the metal respond to the photons in the incoming radiation. The resulting phenomenon propagates in a wave-like manner at the interface between the metal and dielectric, effectively behaving as an electromagnetic wave [53].

After a critical angle of incidence, when the incident light is thought to be fully reflected in the same medium, total internal reflection takes place. There is, nevertheless, a portion of the incident radiation that interacts with the SPs after escaping the medium. This field, known as the evanescent field, extends into the metal from the medium of the lower refractive index. As the wavelength of the input light changes, the evanescent field's penetration depth decreases exponentially in intensity perpendicular to the interface. When the frequencies of the SPs and the evanescent field's photons are equal, the resonant state is created that excites the SPs. This event can be described by the appearance of a dark band in the reflected light. The incident angle at the resonant condition is referred to as the SPR angle or resonance angle, and this phenomenon is known as surface plasmon resonance [53, 54].

The working principle of SPR-based biosensors starts from the preparation of the sensor surface which will be coated with a specific recognition molecule (any antibody, aptamer, etc.) that selectively binds to the antibiotic molecule. Coating of the element ensures the sensor's selectivity toward the target antibiotic. Then the milk sample will be introduced to the surface of the sensor. If the milk sample contains the antibiotic residue, it will attach or bind to the recognition elements on the surface of the sensor. After the binding procedure, local refractive changes occur which induce changes in the angle or wavelength of the surface plasmon resonance is observed, that can be monitored in real-time. The magnitude of the SPR signal correlates with the concentration of antibiotics in the sample. Advanced algorithms process this signal to quantify the amount of antibiotic residue present. The data provides information on the presence and concentration of antibiotic residues.

3.2 Nano-biosensors

3.2.1 Nano Biosensors for Detection of Antibiotic Residues

Biosensors are crucial entities used for the detection and quantification of sundry biologically relevant analytes in many fields of research [55]. Recent advances employ nano-based materials such as nanoparticles, nanotubes, nanorods and nanowires to fabricate nano biosensors for the detection of specific analyte hence the name nano biosensor.

Shrivastav et al. [56] proposed a highly sensitive and selective fiber-based nano sensor for erythromycin detection optics used for milk. Operation capacity of sensor was from 1.62×10^{-3} to 100 μM concentration. LOD for the same is 1.62×10^{-3} μM. Dong et al. [57] developed a sensitive and rapid optical probe (Cerium/zirconium-UiO (Ce/Zr-UiO) 66-based optical biosensor) for the detection of enrofloxacin in 20 milk samples. Outcomes obtained were compared with standard analytical methods HPLC-MS and ELISA. Results determined using MOVE tests were in accordance or similar with the stand methods HPLC-MS and ELISA. The developed probe had a short time for determination (15 min) and was portable.

3.2.2 Quantum Dot-Based Biosensors

Quantum dots (QDs) are recognized as novel fluorescent nanomaterials which consist of inorganic nuclei with organic molecules in the nanoscale range [58]. QDs have been known for the engineering of biosensors for antibiotic detection. Quantum dots are used on the surface of lateral assays, incorporating functionalized recognition elements like antibodies, aptamers, or molecularly imprinted polymers (MIPs) to selectively bind to the antibiotic molecules. The materials used for the purpose are generally silicon, carbon, cadmium, cadmium selenide, cadmium sulfide, etc. [59]. Quantum dots detection methods are prepared using single and/or a combination of various types of material such as Si, ZnS, SiO_2, MIP, etc. Silicon quantum dots have been employed for the controlled release of antibiotics and the breakdown of amoxicillin on their surface through hydrolysis [60]. ZnS/SiO_2/QDs/MIP-based detection procedure was employed for the detection of antibiotic residues in raw milk and milk powder. LOD was noted as 0.81 mg/L. Results depicted material had a long decay time, minimal light scattering, low toxicity, high sensitivity and selectivity, and good binding capacity [61].

3.2.3 Quantum Dot-Based Fluorescence Immunoassay

Fluorescent immunoassays based on Eu-micelles have been used for varied purposes for the rapid detection of fluoroquinolones. Particularly, fluorescent nanoprobes have been extensively used for detection of fluoroquinolone residue in many food products [62]. Quantum dot-based fluorescence immunoassays are fluorescent probes for highly accurate, sensitive and multipurpose detection methods having unique photophysical properties [63].

Wang and research team proposed the use of double-stranded DNA (dsDNA) as an aptamer for antibiotics, one of the DNA strands specifically recognizing chloramphenicol in milk. This method exhibited extremely sensitive behaviour, achieving a detection limit of 2 pg/mL [64]. A novel approach was developed using a quantum dot-based fluorescence immunoassay, where graphene-oxide based QDs (quantum dots) were marked with an aptamer targeting sulfamethazine (SMZ) to detect its residues in milk. This method attained a limit of detection of 5 pg/mL for SMZ [65].

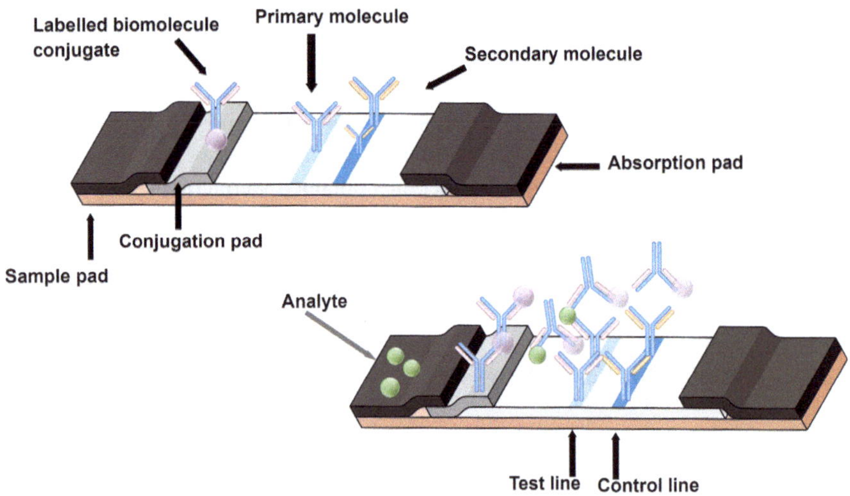

Fig. 2 Quantum dot lateral flow assays and its working principle

Zhu and team developed an innovative near-infrared fluorescence immunoassay that incorporates two types of sensors: quantum dots (QDs) and enzyme-based sensors. The method combines a fluorescence-linked immunosorbent assay, utilizing QD655 and QD605 as probes, with an enzyme-linked immunosorbent assay that uses horseradish peroxidase-labeled secondary antibodies. This integrated assay was designed to detect 11 quinolone and 6 sulfonamide antibiotic residues. Milk samples were tested using this hybrid immunoassay, achieving detection limits of 0.18 ng/mL, 0.17 ng/mL and 7.5 ng/mL for quinolones, sulfonamides, melamine, respectively [66] (Fig. 2).

Ahmed and coworkers fabricated a label-free aptamer-based fluorescence assay for the detection of tetracycline in raw milk. This design approach utilizes aptamer-triggered fluorescence fluctuations of palladium nanoparticles and graphene quantum dots for the rapid finding of tetracycline, achieving a detection limit of 45 ng/mL [67].

3.3 Room-Temperature Phosphorescence

Room-temperature phosphorescence (RTP) is an advanced analytical technique that comprises the emission of light from a substance after it has absorbed photons, with a delay between absorption and emission. The technique operates at room temperature hence it is known to be advantageous. Also, it provides great specificity and high sensitivity.

Traviesa-Alvarez and coworkers prepared metal chelates exhibiting RTP characteristics through the combination of four types of antibiotics from a group of tetracycline (chlortetracycline, doxycycline, tetracycline and oxytetracycline) and europium (III). The chelates are repeatedly injected into the flow and retained in a shared flow cell, which is positioned within the phosphorimeter's cell container. The detection limit was found to be below 0.2 ng/mL in aqueous solutions and 1.92 ng/mL in milk [68]. Albert Redo-Sanchez and colleagues analyzed eleven commonly used antibiotics in animal husbandry. Each antibiotic exhibited a distinct terahertz response within the 0.1 THz to 2 THz range [69]. Ultimately, they were able to differentiate doxycycline and sulfapyridine in powdered milk. Additionally, Yuan et al. [70] conducted a preliminary study on norfloxacin antibiotics. Both quantitative and qualitative analysis of norfloxacin were successfully attained, demonstrating significant potential for food quality detection and inspections. Since the samples tested (such as honey, milk powder, etc.) were in solid form due to water's strong absorption of terahertz waves, the detection limits in these experiments were expressed in terms of weight percent rather than "ng/mL" in liquid samples.

3.4 Surface-Enhanced Raman Scattering (SERS)

SERS is an advanced analytical method that has been used to explore for various many chemicals and biological analyses pertaining to its exclusive property which enhances Raman based signals for analytes [71]. Surface-enhanced Raman spectroscopy (SERS) is a non-invasive technique for analyzing molecular vibrations. This method significantly amplifies Raman signals, increasing their intensity by multiple orders of magnitude. The enhancement occurs when analytes are positioned on or in close proximity to metallic nanostructures [72, 73].

Enhancement mechanisms of SERS: SERS enhancement mechanisms primarily involve two aspects: the electromagnetic mechanism and the chemical mechanism.

Electromagnetic mechanism (EM): Surface plasmon resonance is a phenomenon that takes place when electrons on a metallic nanostructure's surface vibrate in sync with incoming light. This results in a significant amplification of the electromagnetic field, typically within a few nanometers of the nanostructures. When this electromagnetic field is restricted to a confined region on the metal's surface, it is referred to as LSPR (Localised surface plasmon resonance) [72, 74].

Chemical mechanism (CM): This effect relies on the electronic interplay between the metal surface and the molecule adsorbed onto it, including processes like charge transfer. In order to maximise the system's effective polarisability and improve the Raman signal, resonance will be created when the incident photon's energy equals the energy difference of the electron transfer between the metal substrate and the adsorber. Covalently bonded coupled electron clouds, primarily π electrons, are excited electronically within chemical bonds that are generated between analyte molecules and NP surfaces in chemical mechanism [72, 74]. SERS enhancement cab be achieved up to 10^3 fold by chemical mechanism [73].

Under the influence of a certain incident light frequency, the charge transfer between the metal's Fermi level and the highest occupied molecular orbital (HOMO) of the molecule being tested can be produced. Charge transfer between the metal's Fermi level and the molecule's lowest unoccupied molecular orbital (LUMO) can also be induced by choosing a certain incident light irradiation frequency. When the energy of the input photon equals the electron transition energy between the metal substrate and the molecule to be measured, resonance takes place, increasing the adsorbed molecule's effective polarisability causing Raman scattering, which improves the Raman signal [72].

It has been widely utilized for various analytical purposes because of its resistance to water interactions, making it suitable for analyzing water-based systems and enabling in-situ detection of individual marked molecules [75]. Due to this fact, SERS has been known for its efficient detection of target molecules within milk and milk-based food matrices (Fig. 3). During the procedure, the signal of analyte is disrupted by the water molecules which renders SERS to be an important detection method for inspection of other target molecules in any aqueous system. SERS has been known for its unique characteristics of being rapid, highly sensitive, quick and easy to detect. These advantages make it highly competent to even detect a single molecule within any given system [76]. Besides, the other mentioned advantages, SERS can detect molecules within a few minutes without requiring a skilled operator. Moreno et al. [77] employed SERS analytical system consisting of silver nanoparticles (AuNPs) for the detection of sulphonamide in milk. Limit of detection was 8.8 ng/mL. Similarly, Xie et al. [78] found the concentration (0.1 μg/mL) of chloramphenicol classified under amphenicols in milk using SERS having Au-nano substrates.

4 Case Studies

Recently, Zhao et al. [79] developed a portable fiber-optic SPR-based sensor to detect two antibiotic residues namely, enrofloxacin and ciprofloxacin for milk sample. The sensor was created by immobilizing two antibiotic antigens (enrofloxacin and ciprofloxacin) onto the surface of an optical fiber, using bovine serum albumin conjugates through dopamine polymerization. The limit of detection came to be 1.20 ng/mL and 0.81 ng/mL for enrofloxacin and ciprofloxacin, respectively. The sensor's sensitivity came to be 2900 nm/RIU. The sensor prepared had good sensitivity, detection limit and recovery rates. It has a potential in food safety analysis for the detection of antibiotic residues.

Earlier, Gutiérrez et al. [80] developed a nano biosensor for quantification of antibiotic residues (kanamycin, ampicillin, oxytetracycline and sulfadimethoxine) in raw cow milk. These nano biosensors use gold nanoparticles and aptamers which are specifically designed to detect individual antibiotic residue. When antibiotic molecules are present in the milk sample, they interact with the aptamers, causing the random aggregation of gold nanoparticles (AuNPs). Chen and colleagues [81]

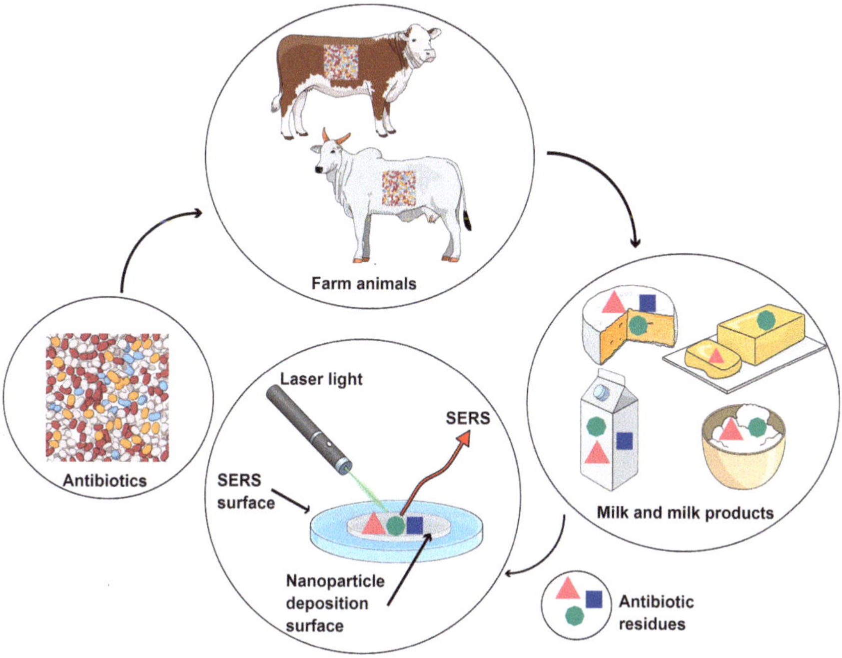

Fig. 3 A schematic illustration of pathways of antibiotic consumption leading to residual accumulation and detection of antibiotic residues using SERS analysis

deposited a permeable aluminum oxide film with adjustable pore sizes, followed by a layer of silver. This resulted in silver being present both on the surface and within the pores of the Al_2O_3 (aluminum oxide). Since the pore size is adjustable, the resonance wavelength of the silver films can also be tuned. By adjusting the pore size of the Al_2O_3, the resonance wavelength can be aligned with the optical spectrum analyzer and light source maximizing the enhancement of Raman signals. The detection limit for chloramphenicol in milk came to be 50 ng/mL, that is significantly higher than the governed regulatory standard i.e. MRL was detected to be 0.3 ng/mL [81].

Khrushchev et al. [82], did a study on determination of trace levels of Fenbendazole in milk, yogurt and curd using SERS with the aim to develop a simple and efficient method for the determination of fenbendazole (FBZ) residues in dairy and fermented dairy products based on SERS. Fenbendazole is a member of the broad-spectrum benzimidazole anthelmintic medication class. It is frequently used to combat all helminth classes, which usually impact various animal species. In order to prevent matrix components from interfering with a subsequent SERS analysis, the sample must be prepared before the analysis. In their study, the materials were purified using an extraction technique to remove any matrix components that interfered. They used 5-Nitrothiabendazole (NTBZ) as an internal standard due to its comparable physicochemical characteristics. The findings demonstrated that the Raman signal might be enhanced by 4.2×107 times due to the process of nanoparticle agglomeration. Firstly, they solved the matrix effect problem by using a two-stage liquid–liquid extraction with heptane and ethyl acetate. Secondly, the determination error caused by analyte losses during sample preparation was greatly reduced by using the internal standard (5-nitrothiabendazole). Thirdly, the results were more reproducible because to the application of the multivariate calibration method (PLS). The limit of detection was 1.8 nM for milk, 8.8 nM for yoghurt, and 13.5 nM for curd. Furthermore, real samples had low LODs ranging from 1.8 to 14.8 nM, which is far lower than the EU-approved maximum residue level of FBZ in dairy products (MRL = 33 nM). This technique differs from the popular liquid chromatography–mass spectrometry (LC–MS) method in that it requires less time to prepare the initial sample and has substantially cheaper reagent and consumable costs.

5 Application of Optical Techniques for Analysis of Antibiotics in Milk and Dairy Products

Applications of optical techniques for analysis of antibiotic residues in milk and dairy products are depicted in Table 3.

6 Comparison of Optical Techniques with Traditional Methods

Traditional Methods

Chromatography was a reliable technique for quantitatively detecting antibiotic residues in milk. It involved separating and analyzing compounds based on their interactions with a stationary phase i.e., high-performance liquid chromatography. Detection limits are typically aligned with established MRLs for specific antibiotics [91]. Culture-based methods involved culturing bacteria from milk samples to identify antibiotic-resistant strains i.e., aminoglycoside, beta-lactam, trimethoprim, fosfomycin and tetracycline. These methods were time-consuming and required specialized facilities [80]. Polymerase chain reaction (PCR) and DNA-based methods also detected previously listed antibiotic resistance genes. These methods were sensitive but may not directly quantify antibiotic residues. Traditional methods are known to be time-demanding and are very less sensitive for the detection of many antibiotic residues.

Optical Techniques

SPR biosensors based on variations in the refractive index near a metal surface to detect binding events. They offer real-time monitoring and high sensitivity. SPR biosensors can provide rapid results and improve analytical performance compared to traditional screening kits for the detection of antibiotics [92]. Colorimetric assays

Table 3 Optical technique used for antibiotic residues detection in milk and dairy products along with their processing parameters and applications

Optical technique	Parameters	Product	Results	Applications	References
SPR-based biosensor	Layers of BK7 prism, 50 mm Ag; 4 mm Si, and monolayer $Ti_3C_2T_x$ nanomaterial	Milk	Highest sensitivity of 350°/RIU	Detection of milk fat concentration and various biosensing applications	[83]
SPR-based biosensor	Silicon substrates were spin-coated with poly (styrene-b-methyl methacrylate) (PS-b-PMMA) thin films to serve as the working electrode; Indium tin oxide-coated glass slides served as the counter electrode; 20 µL of 2.6 nM lipoic acid-functionalized Au nanoparticles	Milk	Sensor's sensitivity was 2900 nm/RIU; detection limits were calculated and came to be 0.81 ng/mL for ciprofloxacin and 1.20 ng/mL for enrofloxacin	Detection of enrofloxacin and ciprofloxacin in milk	[79]
Ultraviolet–visible spectroscopy (UV–Vis)	Silver nanoparticles; a strong peak at 402 nm	Milk, Yogurt	LOD and limit of quantification (LOQ) were found to be 6.2 µM and 18.8 µM (linear range 0.5–10 µM) respectively; the colour change from yellow to orange in the presence of ampicillin	Rapid screening for antibiotics with chromophoric groups	[84]
Fluorescence Spectroscopy	Quantification of antibiotics by mQD-cFIA assay; fluorescence intensity recorded under 370 nm excitation; inverted fluorescence microscope with blue light excitation for 1 s	Milk	Visual detection of penicillin G (PC-G), tetracycline (TC), and streptomycin (SM) in milk; the linear ranges for SM (0.01–25 ng/mL), TC (0.01–25 ng/mL) and PC-G (0.01–10 ng/mL); limit of detection was 5 pg/mL for each residue	Sensitive detection of antibiotics, especially those with low UV–vis absorbance; based on fluorescence of the QD-Ab probes; antibiotic residues can be identified visually and simultaneously	[85]

(continued)

Table 3 (continued)

Optical technique	Parameters	Product	Results	Applications	References
ATR-FTIR (attenuated total reflectance-Fourier transform infrared) spectroscopy	Wavenumber range of 4000–400 cm^{-1} in conjunction with chemometric techniques	Milk, cream	R^2 values for calibration and validation in the 2940–2840 and 1765–1730 cm^{-1} ranges exceeded than 0.92 and 0.94, respectively. Using ATR-FTIR combined with chemometrics, residues get detected in milk at concentrations as low as 1 μg/kg	Rapid meloxicam detection	[86]
NIR (near-infrared spectroscopy)	Eflectance mode in the range of 13.500 to 3.700 cm^{-1} wavenumbers with a Te-InGaAsdetector and 4 cm^{-1} of resolution	Milk	Detection of Enrofloxacin and Oxytetracycline with 100% accuracy; also detect ceftiofur hydrochloride in milk	Real-time monitoring of antibiotic levels in milk production; portable method for antibiotics detection	[87]
Surface-enhanced Raman spectroscopy (SERS)	TiO_2 nanotube array decorated with silver nanoparticles, equipped with a multi-channel air-cooled charge-coupled device (CCD) detector, operating at a wavelength of 532 nm (5 mW)	Milk, Whey	Concentration ranges from 1×10^{-3} to 5×10^{-7}M; SERS intensity and MMT concentration had an excellent linear relationship depicting $R^2 =$ 0.992 value (range of 0.5 to 1000 μM), with a detection limit of 0.11 μM	Highly sensitive detection of antibiotics; especially for residues	[88]

(continued)

Table 3 (continued)

Optical technique	Parameters	Product	Results	Applications	References
Chemiluminescent aptasensor	Light emission; aptamer coated microtiter plates	Milk	Detects antibiotics based on chemical reactions producing light; O-oxytetracycline (0.05 to 5 ng mL^{-1}), T-tetracycline (0.05 to 5 ng mL^{-1}) and K-kanamycin (0.005 to 0.5 ng mL^{-1}); and detection limits for O, T, and K were 0.02, 0.02 and 0.002 ng mL^{-1} (S/N = 3), respectively	Sensitive and specific detection of antibiotics; potential for automation	[89]
Hyperspectral technology and chemometrics	Random forest (RF) and partial least squares-discriminant analysis (PLS-DA) models were established	Milk, Milk powder	PLS-DA with the accuracy of 96.2% and RF with 100% accuracy; The LOD of oxytetracycline, doxycycline, and aureomycin were 1.2×10^{-3}, 1.51×10^{-3}, and 2.44×10^{-3} respectively	Antibiotics (Doxycycline, Chlortetracycline and Oxytetracycline) in milk powder	[90]

utilize the changes in color to indicate the presence of antibiotics. An AuNP-aptamer biosensor visually detected sulfadimethoxine, kanamycin, and adenosine in milk at concentrations of 100–500 ng mL^{-1} [80]. Incorporating nanomaterials i.e., carbon nanomaterials, quantum dots, metal nanomaterials and magnetic nanomaterials enhanced biosensor performance. Nanomaterials are known to have excellent electrical, chemical, optical and mechanical properties. These led to increased sensitivity and selectivity in detecting antibiotics [93, 94].

7 Limitations of Optical Techniques

Despite all the advantages optical methods have over traditional techniques, these still comes with its own limitations. Optical techniques may face interference from other compounds present in food, biological fluids, or environmental samples, leading to false-positive or false-negative results which might hinder its ability to detect sensitive, low levels of residues within the complex food structure. Many antibiotics share similarity among structural features making it challenging for optical methods to distinguish between closely related compounds without additional processing steps. Traditional techniques are cost effective but techniques like SPR, SERS require expensive maintenance and its initial cost is high too. Challenge persists when it comes to quantitative measurement of antibiotic residues. Real time monitoring of antibiotic residues is still under research. While optical techniques offer advantages such as simplicity, rapid analysis, and non-destructive testing, their limitations in sensitivity, selectivity, and cost make them more suitable for screening purposes rather than detailed quantitative analysis.

8 Challenges, Innovations, and Future Directions in Optical Detection for Antibiotics Residues in Milk

Milk is a food system having intricate composition, containing proteins, fats, carbohydrates, minerals, and other components that can interfere with optical detection methods. These matrix effects caused nonspecific binding, quenching, or scattering of the optical signal, leading to inaccuracies. Milk proteins adsorbed onto sensor surfaces and were able to alter the response of the sensor to antibiotics. Precise sample preparation and the use of blocking agents or selective membranes were required to address those problems. The challenge also lies in the potential for cross-reactivity among different antibiotics, where one antibiotic might interfere with the detection of another. A spectral overlap i.e., where the optical signatures of different antibiotics were too similar, can complicate the interpretation of results. Advanced data processing techniques or the use of highly selective receptors are needed to overcome these challenges. The extraction of antibiotics from milk must be efficient enough to prevent the loss of analyte and to avoid introducing contaminants that could interfere with detection. Techniques like solid-phase extraction and liquid–liquid extraction were often used. These technical methods also need to be optimized for each specific antibiotic and sample matrix. Maintaining the stability of the antibiotic during extraction was critical, as some antibiotics are prone to degradation under certain conditions. The development of portable optical devices was essential for the on-farm testing of milk samples. These devices must be robust enough to withstand field conditions, user-friendly for non-specialists, and cost-effective to ensure widespread adoption. The challenge is to maintain sensitivity and specificity in a smaller form factor. These required a novel sensor design and integration.

The use of aptamers-short i.e., selective binding of single-stranded DNA/RNA molecules to specific target molecules, offers a highly selective approach for antibiotic detection. Molecular imprinting i.e., polymers were created with cavities shaped like the specific antibiotic, provided another route to high selectivity. These receptors can reduce the risk of cross-reactivity. Plasmonic nanoparticles i.e., gold and silver nanoparticles, can be used to boost optical signals through surface-enhanced Raman scattering (SERS) or localized surface plasmon resonance (LSPR). These techniques amplify the signal from a target molecule, enabling the detection of antibiotics at extremely low concentrations. To mitigate matrix effects, the use of coatings or blocking agents on sensor surfaces can prevent nonspecific adsorption of milk components. Microfluidic platforms offer precise control over fluid handling, which is beneficial for reducing sample and reagent volumes, minimizing waste, and improving reaction kinetics. Microfluidic devices can enhance sensitivity and enable high-throughput screening of antibiotics in milk when combined with optical sensors. Microfluidic devices can be suitable for on-site testing.

Advanced nanomaterials such as graphene, metal–organic frameworks, and quantum dots hold great promise for improving the sensitivity and selectivity of optical sensors. These materials have unique optical properties i.e., varying fluorescence or enhanced Raman scattering. The complexity of optical spectra often requires pioneering data analysis techniques. An expert system or machine learning algorithms can be trained to recognize patterns in the data that correspond to the presence of specific antibiotics. The development of implantable or wearable optical biosensors for continuous monitoring of antibiotic levels in dairy animals represents a significant advance in real-time data collection. These wearable devices could provide continuous feedback on antibiotic levels, enabling more precise dosing and reducing the risk of antibiotic residues in milk.

9 Conclusion

Antibiotic residues have been a major concern in the present scenario and are being well-researched due to their hazardous effects transferred from consumable products to living beings. Optical methods for detecting antibiotics offer key advantages like rapid response and high sensitivity. However, improvements in sensitivity, portability, regeneration efficiency, and cost-efficient are needed to further enhance their working ability and broaden their application scope. More innovations are needed for quick and sensitive detection methods.

References

1. National Institutes of Health (NIH), Office of Dietary Supplements. Calcium: Fact Sheet for Consumers (2020). https://ods.od.nih.gov/factsheets/Calcium-Consumer/
2. K.F. Michaelsen, A. Larnkjaer, Milk and milk products: good or bad for human health? An assessment of the totality of scientific evidence. Food Nutr. Res. **55**(1) (2011). https://doi.org/10.3402/fnr.v55i0.5153
3. B. Kukanich, R. Gehring, A.I. Webb, A.L. Craigmill, Use of antimicrobials in food animal species. J. Am. Vet. Med. Assoc. **252**(6), 701–711 (2018)
4. K. Tiseo, L. Huber, M. Gilbert, T.P. Robinson, T.P. Van Boeckel, Global trends in antimicrobial use in food animals from 2017 to 2030. Antibiotics **9**(12), 918 (2020)
5. C. Gonzalo, J.A. Carriedo, M.C. García-Jimeno, M. Pérez-Bilbao, L.F. de la Fuente, Factors influencing variation of bulk milk antibiotic residue occurrence, somatic cell count, and total bacterial count in dairy sheep flocks. J. Dairy Sci. **93**, 1587–1595 (2010)
6. Food and Drug Administration, Department of Health, and Human Services, Animal drugs, feeds, and related products: tolerances for residues of new animal drugs in food, **21**(6) (2023). Accessed on 05 April 2024 https://www.accessdata.fda.gov/scripts/cdrh/cfdocs/cfcfr/CFRSearch.cfm?CFRPart=556&showFR=1
7. S. Sachi, J. Ferdous, M.H. Sikder, S.A.K. Hussani, Antibiotic residues in milk: past, present, and future. J. Adv. Vet. Animal Res. **6**(3), 315 (2019)
8. A. Rama, L. Lucatello, C. Benetti, G. Galina, D. Bajraktari, Assessment of antibacterial drug residues in milk for consumption in Kosovo. J. Food Drug Anal. **25**(3), 525–532 (2017)
9. P. Moudgil, J.S. Bedi, R.S. Aulakh, J.P.S. Gill, Antibiotic residues and mycotoxins in raw milk in Punjab (India): a rising concern for food safety. J. Food Sci. Technol. **56**, 5146–5151 (2019)
10. A. Gallo, R. Landi, V. Rubino, A. Di Cerbo, A. Giovazzino, A.T. Palatucci, G. Terrazzano, Oxytetracycline induces DNA damage and epigenetic changes: a possible risk for human and animal health? PeerJ **5**, e3236 (2017)
11. M. Rong, Y. Huang, C. Lin, L. Lai, Y. Wu, L. Niu, Recent Advances in optical sensing for tetracycline antibiotics. TrAC Trends Anal Chem. 117839 (2024)
12. Q. Wang, W.M. Zhao, Optical methods of antibiotic residues detections: a comprehensive review. Sens. Actuators: B. Chem. **269**, 238–256 (2018). https://doi.org/10.1016/j.snb.2018.04.097
13. P. Navrátilová, J. Vyhnálková, L. Vorlová, Detection of fluoroquinolone residues in milk using *Yersinia* spp. strains: towards better sensitivity for flumequine determination. J. Vet. Res. **61**(4), 467–472 (2017)
14. M. Jospe-Kaufman, L. Siomin, M. Fridman, The relationship between the structure and toxicity of aminoglycoside antibiotics. Bioorg. Med. Chem. Lett. **30**(13), 127218 (2020)
15. S.W. Kim, J.S. Lee, S.B. Park, A.R. Lee, J.W. Jung, J.H. Chun, T.S. Jung, The importance of porins and β-lactamase in outer membrane vesicles on the hydrolysis of β-lactam antibiotics. Int. J. Mol. Sci. **21**(8), 2822 (2020)
16. W. Simonson, Antibiotic stewardship: revisiting quinolone antibiotics. Geriatric Nursing (New York, NY) **38**(2), 152–153 (2017)
17. J.P. Ferguson, G.A. Baxter, J.D.G. McEvoy, S. Stead, E. Rawlings, M. Sharman, Detection of streptomycin and dihydrostreptomycin residues in milk, honey and meat samples using an optical biosensor. Analyst **127**(7), 951–956 (2002)

18. A. Rahman, P. Paul, M.R. Sarkar, K.M.Y.K. Sikdar, I.Z. Esti, N.M. Abid, A.B.M. Faroque, Antibiotic residues in pasteurised and Raw Cow's milk in Dhaka, Bangladesh. Food Addit. Contam. Part B (2024). https://doi.org/10.1080/19393210.2024.2335234
19. A. Kumar, A.K. Panda, N. Sharma, Determination of antibiotic residues in bovine milk by HPLC-DAD and assessment of human health risks in Northwestern Himalayan region, India. J. Food Sci. Technol. **59**(1), 95–104 (2022)
20. P. Quintanilla, M.C. Beltrán, M.P. Molina, I. Escriche, Enrofloxacin treatment on dairy goats: presence of antibiotic in milk and impact of residue on technological process and characteristics of mature cheese. Food Control **123**, 107762 (2021)
21. N.M. Patel, R. Kumar, C.V. Savalia, D.N. Desai, I.H. Kalyani, Dietary exposure and risk assessment of antibiotics residues in marketed bovine raw milk. J. Entomol. Zool. Stud. **8**(4), 1823–1827 (2020)
22. M.A. Hebbal, C. Latha, K.V. Menon, J. Deepa, Occurrence of oxytetracycline residues in milk samples from Palakkad, Kerala, India. Vet. World **13**(6), 1056 (2020)
23. N.P. Kumarswamy, C. Latha, K.M. Vrinda, C. Sethukekshmi, K.A. Mercy, Detection of antibiotic residues in raw cow milk in Thrissur, India. Pharma Innov. J. **7**(8), 452–454 (2018)
24. R.K. Nirala, K. Anjana, K.G. Mandal, C. Jayachandran, Persistence of antibiotic residue in milk under region of Bihar, India. Int. J. Curr. Microbiol. App. Sci. **6**(3), 2296–2299 (2017)
25. I.O. Olatoye, O.F. Daniel, S.A. Ishola, Screening of antibiotics and chemical analysis of penicillin residue in fresh milk and traditional dairy products in Oyo state, Nigeria. Vet. World **9**(9), 948 (2016)
26. S. Chowdhury, M.M. Hassan, M. Alam, S. Sattar, M.S. Bari, A.K.M. Saifuddin, M.A. Hoque, Antibiotic residues in milk and eggs of commercial and local farms at Chittagong, Bangladesh. Vet. World **8**(4), 467 (2015)
27. N. Zheng, J. Wang, R. Han, X. Xu, Y. Zhen, X. Qu, Z. Yu, Occurrence of several main antibiotic residues in raw milk in 10 provinces of China. Food Addit. Contam. Part B **6**(2), 84–89 (2013)
28. M.S. Rahman, M.M. Hassan, S. Chowdhury, Determination of antibiotic residues in milk and assessment of human health risk in Bangladesh. Heliyon **7**(8) (2021)
29. J.M. Diserens, A. Beck Henzelin, M.H. Le Breton, M.C. Savoy Perroud, Current situation & compilation of commercially available screening methods for the detection of inhibitors/antibiotic residues in milk. Fil-DF Bull. Federation Internationale de Laiterie International Dairy Federation **442**, 2 (2010)
30. M.K. Elbayoumy, A.M. Allam, A.A. Ghazy, S.M. Nasr, Advances in controlling bacterial mastitis in dairy cows. Egypt. J. Vet. Sci. **55**(1), 1–21 (2024)
31. M. Eskandari, Z. Abdolmaleki, F. Moosakhani, M.A. Eslampour, Antimicrobial efficacy of cloxacillin-loaded chitosan nanoparticles against staphylococcus aureus biofilms in subclinical mastitis. Curr. Microbiol. **82**(2), 65 (2025)
32. Y.T. Grohn, D.J. Wilson, R.N. González, J.A. Hertl, H. Schulte, G.J. Bennett, P.M. Sears, Effect of pathogen-specific clinical mastitis on milk yield in dairy cows. J. Dairy Sci. **88**(12), 4110–4124 (2005)
33. S.P. Oliver, S.E. Murinda, B.M. Jayarao, Impact of antibiotic use in adult dairy cows on antimicrobial resistance of veterinary and human pathogens: a comprehensive review. Foodborne Pathog. Dis. **8**(3), 337–355 (2011)
34. G. Kebede, T. Zenebe, H. Disassa, T. Tolosa, Review on detection of antimicrobial residues in raw bulk milk in dairy farms. Afr. J. Basic Appl. Sci. **6**(4), 87–97 (2014)
35. S. Máttar, A. Calderón, D. Sotelo, M. Sierra, G. Tordecilla, Detección de antibióticos en leches: un problema de salud pública. Revista de Salud Pública **11**, 579–590 (2009)
36. L. Zhou, Z. Luoreng, X. Wang, Identification of potential key circular RNAs associated with *Escherichia coli*-infected bovine mastitis using RNA-sequencing: preliminary study results. Vet. Res. Commun. **49**(1), 1–12 (2025)
37. A.E. Hesham, R.N.A. El-Latif, H.M. Zain, Detection of β-lactam antibiotic residues in bovine milk in Egypt and Sudan. Int. Food Res. J. **21**(2), 803–807 (2014)
38. R. Khosrokhavar, M. Jamal, S. Shahram, B. Jannat, Detection of oxytetracycline residue in infant formulary high-performance liquid chromatography. Iran. J. Pharm. Res. 1–32 (2011)

39. J. Li, Y. Zhen, Impact of tetracycline on microbial community structure and function in soils with different properties. J. Environ. Sci. **47**, 89–97 (2016)
40. I. Cho, S. Yamanishi, L. Cox, B.A. Methé, J. Zavadil, K. Li, M.J. Blaser, Antibiotics in early life alter the murine colonic microbiome and adiposity. Nature **488**(7413), 621–626 (2012)
41. R.C. Chandan, Dairy–fermented products. Food Process. Principles Appl. 405–436 (2014)
42. T. Varzakas, Microbiological safety systems for dairy processing. Dairy Microbiol. Biochem. Recent Dev. **356** (2014)
43. F. Prestinaci, P. Pezzotti, A. Pantosti, Antimicrobial resistance: a global multifaceted phenomenon. Pathog. Glob. Health **109**(7), 309–318 (2015)
44. T. Beyene, Veterinary drug residues in food-animal products: its risk factors and potential effects on public health. J. Vet. Sci. Technol. **7**(1), 1–7 (2016)
45. World Health Organization, *Critically Important Antimicrobials for Human Medicine (5th revision)* (World Health Organization, Geneva, 2014)
46. B.M. Marshall, S.B. Levy, Food animals and antimicrobials: impacts on human health. Clin. Microbiol. Rev. **24**(4), 718–733 (2011)
47. E. Kaltenthaler, P. Shackley, Stevens-Johnson syndrome/toxic epidermal necrolysis and erythema multiforme in children: a review of 10 years' experience. Int. J. Clin. Pract. **54**(4), 240–243 (2000)
48. G. Desa, Antibiotic residues in cow milk and its public health importance review. Adv. Biol. Res. **13**(4), 164–172 (2019)
49. B. Abad, D.A. Borca-Tasciuc, M.S. Martin-Gonzalez, Non-contact methods for thermal properties measurement. Renew. Sustain. Energy Rev. **76**, 1348–1370 (2017)
50. S. Roose, Y. Stockman, P. Rochus, T. Kuhn, M. Lang, H. Baier, G. Casarosa, Optical methods for non-contact measurements of membranes. Acta Astronaut. **65**(9–10), 1317–1329 (2009)
51. G. Okoro, S. Husain, M. Saukani, C. Mutalik, S. Yougbaré, Y.C. Hsiao, T.R. Kuo, Emerging trends in nanomaterials for photosynthetic biohybrid systems. ACS Mater. Lett. **5**(1), 95–115 (2022)
52. A. Philip, A.R. Kumar, The performance enhancement of surface plasmon resonance optical sensors using nanomaterials: a review. Coord. Chem. Rev. **458**, 214424 (2022)
53. N. Ravindran, S. Kumar, M. CA, S.N. Thirunavookarasu, S. CK, Recent advances in Surface Plasmon Resonance (SPR) biosensors for food analysis: a review. Crit. Rev. Food Sci. Nutr. **63**(8), 1055–1077 (2023)
54. Y. Yanase, T. Hiragun, K. Ishii, T. Kawaguchi, T. Yanase, M. Kawai, M. Hide, Surface plasmon resonance for cell-based clinical diagnosis. Sensors **14**(3), 4948–4959 (2014)
55. P. Bhattarai, S. Hameed, Basics of biosensors and nanobiosensors. Nanobiosens. Des. Appl. 1–22 (2020)
56. A.M. Shrivastav, S.P. Usha, B.D. Gupta, Highly sensitive and selective erythromycin nanosensor employing fiber optic SPR/ERY imprinted nanostructure: application in milk and honey. Biosens. Bioelectron. **90**, 516–524 (2017)
57. B. Dong, K. Hu, Y. Mao, K. Wen, Z. Wang, H. Qu, L. Zheng, A nanomaterial-independent and fluorescent immunoassay based on Eu-micelles for rapid and sensitive detection of fluoroquinolones in chicken. Food Chem. **459**, 140419 (2024)
58. P. Li, S. Liu, W. Cao, G. Zhang, X. Yang, X. Gong, X. Xing, Low-toxicity carbon quantum dots derived from gentamicin sulfate to combat antibiotic resistance and eradicate mature biofilms. Chem. Commun. **56**(15), 2316–2319 (2020)
59. J.W. Zhou, X.M. Zou, S.H. Song, G.H. Chen, Quantum dots applied to methodology on detection of pesticide and veterinary drug residues. J. Agric. Food Chem. **66**(6), 1307–1319 (2018)
60. A. Das, P.T. Snee, Synthetic developments of nontoxic quantum dots. Chem. Phys. Chem. **17**(5), 598–617 (2016)
61. S. Chen, Y. Li, S. Wu, X. Jiang, H. Yang, X. Su, Y. Yang, A phosphorescent probe for cephalexin consisting of mesoporous thioglycolic acid-modified Mn: ZnS quantum dots coated with a molecularly imprinted polymer. Microchim. Acta **187**, 1–10 (2020)

62. Y. Dong, Y. Zhang, P. Liu, S. Zhu, X. Peng, X. Hu, Y. Chen, A metal-organic framework signaling probe-mediated immunosensor for the economical and rapid determination of enrofloxacin in milk. Food Chem. **449**, 139050 (2024)
63. Y. Lv, J. Fan, M. Zhao, R. Wu, L.S. Li, Recent advances in quantum dot-based fluorescence-linked immunosorbent assays. Nanoscale **15**(12), 5560–5578 (2023)
64. Y. Wang, N. Gan, T. Li, Y. Cao, F. Hu, Y. Chen, A novel aptamer–quantum dot fluorescence probe for specific detection of antibiotic residues in milk. Anal. Methods **8**(15), 3006–3013 (2016)
65. Y. He, B. Zhang, Z. Fan, Aptamer based fluorometric sulfamethazine assay based on the use of graphene oxide quantum dots. Microchim. Acta **185**, 1–8 (2018)
66. K. Zhu, J. Li, Z. Wang, H. Jiang, R.C. Beier, F. Xu, S. Ding, Simultaneous detection of multiple chemical residues in milk using broad-specificity antibodies in a hybrid immunosorbent assay. Biosens. Bioelectron. **26**(5), 2716–2719 (2011)
67. S.R. Ahmed, S. Kumar, G.A. Ortega, S. Srinivasan, A.R. Rajabzadeh, Target specific aptamer-induced self-assembly of fluorescent graphene quantum dots on palladium nanoparticles for sensitive detection of tetracycline in raw milk. Food Chem. **346**, 128893 (2021)
68. J.M. Traviesa-Alvarez, J.M. Costa-Fernández, R. Pereiro, A. Sanzmedel, Direct screening of tetracyclines in water and bovine milk using room temperature phosphorescence detection. Anal. Chim. Acta **589**, 51–58 (2007). https://doi.org/10.1016/j.aca.2007.02.063
69. A. Redo-Sanchez, G. Salvatella, R. Galceran, E. Roldós, J.A. Garcia-Reguero, M. Castellari, J. Tejada, Assessment of terahertz spectroscopy to detect antibiotic residues in food and feed matrices. Analyst **136**(8), 1733–1738 (2011)
70. Y. Yuan, D. Cao, Y. Zhang, J. Ma, J. Qi, Q. Wang, ... & G.F. Gao, Cryo-EM structures of MERS-CoV and SARS-CoV spike glycoproteins reveal the dynamic receptor binding domains. Nat. Commun. **8**(1), 15092 (2017)
71. M. Girmatsion, A. Mahmud, B. Abraha, Y. Xie, Y. Cheng, H. Yu, H. Qian, Rapid detection of antibiotic residues in animal products using surface-enhanced Raman Spectroscopy: a review. Food Control **126**, 108019 (2021)
72. W. Jiang, Q. Wang, K. Chang, Y. Zhao, Surface-enhanced Raman spectroscopy substrates for monitoring antibiotics in dairy products: mechanisms, advances, and prospects. Compr. Rev. Food Sci. Food Saf. **23**(6), e70024 (2024)
73. H.K. Lee, Y.H. Lee, C.S.L. Koh, G.C. Phan-Quang, X. Han, C.L. Lay, X.Y. Ling, Designing surface-enhanced Raman scattering (SERS) platforms beyond hotspot engineering: emerging opportunities in analyte manipulations and hybrid materials. Chem. Soc. Rev. **48**(3), 731–756 (2019)
74. N.E. Dina, M.A. Tahir, S.Z. Bajwa, I. Amin, V.K. Valev, L. Zhang, SERS-based antibiotic susceptibility testing: towards point-of-care clinical diagnosis. Biosens. Bioelectron. **219**, 114843 (2023)
75. V. Balan, C.T. Mihai, F.D. Cojocaru, C.M. Uritu, G. Dodi, D. Botezat, I. Gardikiotis, Vibrational spectroscopy fingerprinting in medicine: from molecular to clinical practice. Materials **12**(18), 2884 (2019)
76. R.F. Peters, L. Gutierrez-Rivera, S.K. Dew, M. Stepanova, Surface enhanced Raman spectroscopy detection of biomolecules using EBL fabricated nanostructured substrates. JoVE (J. Visualized Exp.) **97**, e52712 (2015)
77. V. Moreno, A. Adnane, R. Salghi, M. Zougagh, Á. Ríos, Nanostructured hybrid surface enhancement Raman scattering substrate for the rapid determination of sulfapyridine in milk samples. Talanta **194**, 357–362 (2019)
78. Y. Xie, M. Zhao, Q. Hu, Y. Cheng, Y. Guo, H. Qian, W. Yao, Selective detection of chloramphenicol in milk based on a molecularly imprinted polymer–surface-enhanced Raman spectroscopic nanosensor. J. Raman Spectrosc. **48**(2), 204–210 (2017)
79. Z. Zhao, H. Yin, J. Xiao, M. Cui, R. Huang, R. Su, Efficient sequential detection of two antibiotics using a fiber-optic surface Plasmon resonance sensor. Sensors **24**(7), 2126 (2024). https://doi.org/10.3390/s24072126

80. P. Gutiérrez, S.E. Godoy, S. Torres, P. Oyarzún, I. Sanhueza, V. Díaz-García, B. Contreras-Trigo, P. Coelho, Improved antibiotic detection in raw milk using machine learning tools over the absorption spectra of a problem-specific nanobiosensor. Sensors **20**(16), 4552 (2020). https://doi.org/10.3390/s20164552
81. J. Chen, S. Feng, F. Gao, E. Grant, J. Xu, S. Wang, X. Lu, Fabrication of SERS-active substrates using silver nanofilm-coated porous anodic aluminum oxide for detection of antibiotics. J. Food Sci. **80**(4), N834–N840 (2015)
82. A.Y. Khrushchev, A.Y. Gulyaeva, V.O. Bondarenko, Determination of trace levels of fenbendazole in milk, yogurt and curd using surface enhanced Raman spectroscopy (SERS). Anal. Lett. **58**(3), 386–400 (2024)
83. A.H.M. Almawgani, M.G. Daher, S.A. Taya, M. Mashagbeh, I. Colak, Optical detection of fat concentration in milk using MXENE-based surface Plasmon resonance structure. Biosensors **12**(7), 535 (2022). https://doi.org/10.3390/bios12070535
84. A. Mishra, E.A. Kukreja, R.N. Pudake, R. Kumar, M.P. Singh, L. Yadav, I. Pandey, F.A. Sheikh, R.M. Tripathi, Label-free selective and sensitive colorimetric detection of ampicillin in milk and water using silver nanoparticles. J. Food Compos. Anal. **119**, 105256 (2023). https://doi.org/10.1016/j.jfca.2023.105256
85. E. Song, M. Yu, Y. Wang, W. Hu, D. Cheng, M.T. Swihart, Y. Song, Multi-color quantum dot-based fluorescence immunoassay array for simultaneous visual detection of multiple antibiotic residues in milk. Biosens. Bioelectron. **72**, 320–325 (2015). https://doi.org/10.1016/j.bios.2015.05.018
86. R. Saji, K. Gandhi, R. Sharma, H. Raghu, Detection of meloxicam residues in milk using ATR-FTIR spectroscopy coupled with chemometrics. Food Control **163**, 110491 (2024). https://doi.org/10.1016/j.foodcont.2024.110491
87. L. Da Conceição Luiz, M.J.V. Bell, R.A. Da Rocha, N.L. Leal, V. De Carvalho Dos Anjos, Detection of veterinary antimicrobial residues in milk through near-infrared absorption spectroscopy. J. Spectrosc. 1–6 (2018). https://doi.org/10.1155/2018/5152832
88. M. Jing, H. Zhang, M. Li, Z. Mao, X. Shi, Silver nanoparticle-decorated TiO_2 nanotube array for solid-phase microextraction and SERS detection of antibiotic residue in milk. Spectrochim. Acta Part A Mol. Biomol. Spectrosc. **255**, 119652 (2021). https://doi.org/10.1016/j.saa.2021.119652
89. L. Hao, H. Gu, N. Duan, S. Wu, Z. Wang, A chemiluminescent aptasensor for simultaneous detection of three antibiotics in milk. Anal. Methods **8**(44), 7929–7936 (2016). https://doi.org/10.1039/c6ay02304e
90. J. Hu, Z. Xu, M. Li, Y. He, Y. Liu, Discriminant analysis and quantitative study of antibiotics in infant milk powder based on hyperspectral detection. Vib. Spectrosc. **114**, 103244 (2021). https://doi.org/10.1016/j.vibspec.2021.103244
91. D. Yamin, V. Uskoković, A. Wakil, M. Goni, S. Shamsuddin, F. Mustafa, W. Alfouzan, M. Alissa, A. Alshengeti, R. Almaghrabi, M. Fares, M. Garout, N.A. Kaabi, A. Alshehri, H. Ali, A. Rabaan, F. Aldubisi, C. Yean, N. Yusof, Current and future technologies for the detection of antibiotic-resistant bacteria. Diagnostics **13**(20), 3246 (2023). https://doi.org/10.3390/diagnostics13203246
92. K. Kivirand, M. Kagan, T. Rinken, *Biosensors for the Detection of Antibiotic Residues in Milk* (InTech eBooks, 2015). https://doi.org/10.5772/60464
93. M. Chudasama, J. Goyary, Nanostructured materials in food science: current progress and future prospects. Next Mater. (2024). https://doi.org/10.1016/j.nxmate.2024.100206
94. S. Seth, P. Rathinasabapathi, A short review on detection of antibiotics in milk using nanomaterial-based biosensor. Food Anal. Methods **15**(8), 2181–2192 (2022). https://doi.org/10.1007/s12161-022-02291-6

Recent Advances in Nanomaterial-Based Optical Probes for the Detection of Food Adulterants

Shalini Singh and Umesh K. Tiwari

Abstract Food adulteration is defined as the intentional alteration of food composition for economic gain or deceptive practices. The detection of food adulterants poses a significant challenge to food safety and consumer trust. The advancement in nanomaterial-based optical probes offers promising solutions to enhance the sensitivity, and selectivity of detecting various adulterants in food products. This book chapter provides an overview of the latest developments in nanomaterial-enabled fluorescence, SPR, and SERS-based optical sensing techniques for the detection of food adulterants. Researchers have developed innovative sensing platforms capable of detecting adulterants at trace levels by leveraging the unique optical properties of nanomaterials, such as plasmonic nanoparticles, quantum dots, and carbon-based nanomaterials. This chapter discusses the principles underlying nanomaterial-based optical probes and the recent advancements in technologies to detect the adulteration to highlight the design and fabrication and their potential applications in ensuring food authenticity and its safety.

Keywords Food adulteration · Fluorescence · SPR · SERS · Nanomaterials

1 Introduction

Food adulteration refers to the degradation of the quality of food either by the adding, replacing or substituting of foreign ingredient. It involves illegal practices that are prohibited by food laws and regulations. It severely impacts the consumer health and safety [53]. For instance, milk is one of the most adulterated foods with water, detergents such sodium hydroxide (caustic soda), hydrated lime, ammonium sulphate, and urea to increase volume and enhance its viscosity, all these results in serious health risks to kidney and digestive system. In addition, vegetable oils are mixed

S. Singh · U. K. Tiwari (✉)
CSIR-Central Scientific Instruments Organisation, Chandigarh, India
e-mail: umeshtiwari@csio.res.in

Academy of Scientific and Innovative Research (AcSIR), Ghaziabad, India

R. Biswas (ed.), *Optical Techniques for Assessing Food Adulterants*, Progress in Optical Science and Photonics 35, https://doi.org/10.1007/978-981-96-8142-6_6

with cheaper, non-edible oils like raw cottonseed oil, which is associated with infertility and liver damage. Similarly, honey is often diluted with sugar syrups or high fructose corn syrup, misleading consumers and contributing to obesity and diabetes. Spices like turmeric and chili powder are frequently mixed with artificial colors and brick powder to improve their visual appeal, potentially leading to severe allergic reactions and chronic illnesses due to the ingestion of toxic substances. Saffron is an expensive spice that could commonly adulterated with lower-quality substances such as artificial colors or other plant derivatives with similar chemical composition like sumac powder, calendula, rubia powder, pomegranate flower, and curcumin, Safflower stamens etc. Phenyl-azoic Sudan dyes are a class of synthetic colorants commonly employed in the chemical industry to tint hydrocarbon solvents, oils, and waxes. Research has identified the Sudan dyes as human carcinogens and mutagens, leading to restrictions on their use as food additives by the Food Standards Agency and the European Union [12]. The adulteration of fruits and vegetables with harmful pesticides and illegal dyes to provide rich colors beyond permissible limits is a major concern, leading to long-term health problems including cancer and neurological disorders [22]. The health implications of food adulteration are profound and alarming. Therefore, the food industry urgently needs stringent regulatory measures and consumer awareness to combat food adulteration and ensure the safety of the food supply. Governments and health organizations have established guidelines and standards to combat food adulteration. From Indian context, to regulate the standard quality, manufacturing, packaging, storage, distribution, sale and import of the food, Food Safety and Standards Authority of India (FSSAI) was established under the Food Safety and Standard Act 2006. The main aim of FSSAI is to ensure the availability of safe and wholesome food for human consumption (Fig. 1).

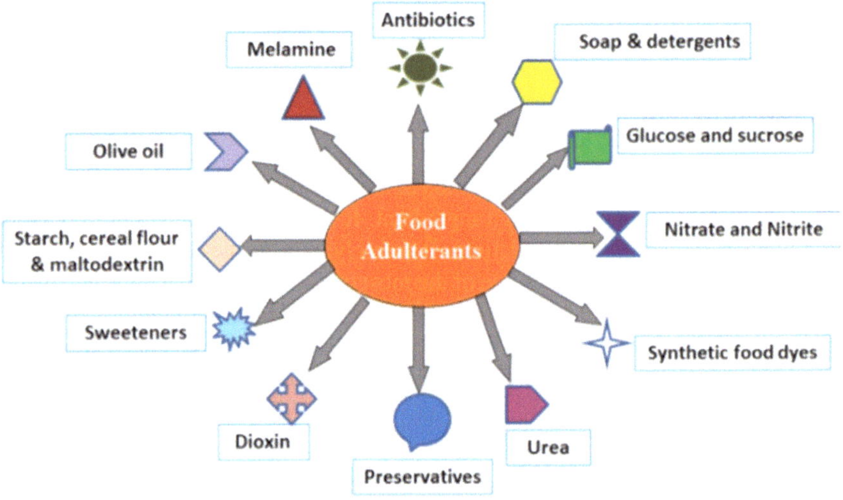

Fig. 1 Types of food adulterants [79]

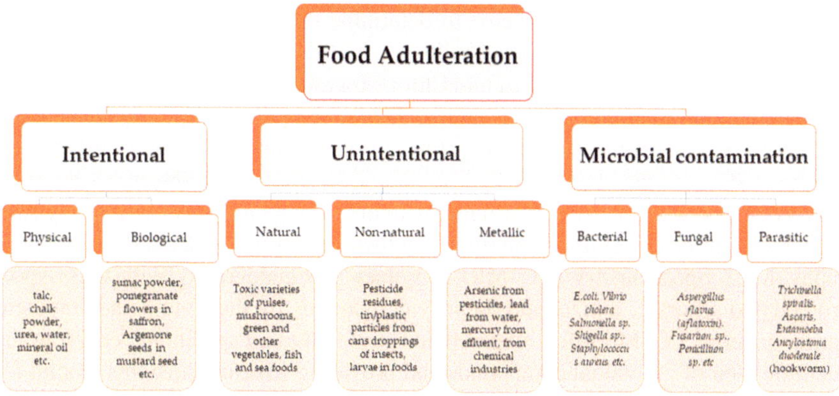

Fig. 2 Classification of food adulteration

2 Classification of Food Adulteration

Intentional adulteration is the phenomenon of deliberately adding food adulterants to products to increase quantity, alter appearance, improve texture, or reduce production costs, often at the expense of quality, safety, and nutritional value. These adulterants can be harmful to human health and are usually added to deceive consumers for economic gain. Unintentional or incidental adulteration is when food adulterants can unintentionally contaminate food products during production, processing, packaging, or transportation. These adulterants are not added deliberately but result from environmental contamination, improper handling, or inadequate hygiene practices. Microbial contamination refers to the unintended presence of harmful microorganisms, such as bacteria, viruses, fungi, or parasites, in food products. This type of contamination can occur at any stage of the food supply chain, including production, processing, storage, or distribution, which can cause foodborne illness [7]. The subclass of the above classification is mentioned in the Fig. 2.

3 Conventional and Advanced Techniques for Detection of Food Adulterants

Analytical tests play a crucial role in ensuring the safety and authenticity of food products, particularly in detecting adulteration with food dyes and other similar chemical structural components. Spectrophotometry utilizes specific wavelengths of light to measure compound quantities based on their absorption characteristics, making it efficient for analyzing highly pigmented substances. However, its drawback lies in limited specificity, as it cannot reliably distinguish between synthetic and natural dyes [57]. Thin Layer Chromatography (TLC) operates by separating compounds on a silica gel plate using solvents. It provides distinct patterns for each compound, enabling comparison with standards to determine their presence in food products [9]. However, this technique faces challenges in differentiating structurally similar compounds. High-Performance Liquid Chromatography with UV detection (HPLC–UV) extends chromatographic separation capabilities by detecting compounds based on their UV absorption after passing through a column. It excels in quantifying synthetic dyes at low levels and measuring their concentrations precisely [61]. Likewise, TLC and HPLC–UV struggle with discerning closely related compounds that share similar structural properties. Another powerful tool, Liquid Chromatography–Mass Spectrometry (LC–MS) combines chromatography with mass spectrometry to separate, ionize, and identify compounds by their mass and fragmentation patterns. LC–MS offers unparalleled sensitivity and specificity, enabling the identification of individual compounds and providing structural insights crucial for distinguishing between natural and synthetic dyes. Despite its effectiveness, LC–MS requires advanced instrumentation and expertise, making it less accessible for routine food testing than simpler techniques like spectrophotometry and TLC [78]. These analytical methods offer unique advantages for detecting food adulteration but also present challenges that impact its applicability in food safety and quality assurance contexts. The above-discussed conventional techniques necessitate large quantities of solvents and reagents and extensive sample preparation, which can alter the sample during the detection process. Furthermore, these methods are often intricate, time-consuming, labor-intensive, and unsuitable for large-scale applications. In contrast, the advanced detection methods offer distinct advantages as simple, rapid, and non-invasive methods for the identification of adulterants. This chapter explores these innovative analytical techniques in detail, highlighting their principles, applications, and performance through examples. A comparative summary of their advantages and limitations is presented in Table 1, offering insights into the suitability of these methods for various scenarios discussed throughout the chapter.

4 Nanomaterial-Based Optical Probes for the Detection of Food Adulterants

4.1 Fluorescence-Based Optical Probes

Among optical-based detection methods, fluorescence sensors are attractive due to their advantageous characteristics, including high sensitivity, multiplexing capability, low limits of detection, simple operation, and rapid response times, making them well-suited for food contaminants. The rapid progress in nanotechnology and materials science has led to significant advancements and breakthroughs in monitoring of food and its safety [73]. Notably, innovative materials, i.e. metal nanoclusters, functional nanomaterials, quantum dots, carbon dots, fluorescence nanoparticles, and Porous coordination polymers, have been intensively explored for the development of high-performance food sensors [36, 112]. Therefore, the integration of these

Table 1 Different analytical techniques with their advantages and disadvantages

S. no	Detection techniques	Advantages	Disadvantages
1	Fluorescence-based probes	High sensitivity and specificity, Wide dynamic range, rapid and real-time detection, minimal sample preparation	Potential for photobleaching, possible interference from autofluorescence of food matrix, requires careful calibration and standardization, can be expensive due to the cost of fluorophores
2	SPR-based probes	Real-time and label-free detection, High specificity due to biomolecular interactions quantitative analysis possible, minimal sample preparation	High cost of SPR equipment, requires a clean and stable sample surface, sensitivity to temperature and refractive index changes, limited multiplexing capabilities
3	Raman spectroscopy probes	Non-destructive and rapid, minimal sample preparation, Provides molecular fingerprinting, suitable for complex food matrices	Low sensitivity compared to SERS, fluorescence background can interfere with Raman signals, requires expensive and sophisticated equipment, not suitable for trace-level detection without enhancement
4	SERS-based probes	Ultra-high sensitivity, capable of detecting low concentrations, enhances weak Raman signals from non-specific adsorption, suitable for trace analysis in complex matrices	Requires careful preparation of nanostructures, Reproducibility issues due to variability in nanostructures, potential interference from non-specific adsorption, high cost of materials and equipment

nanomaterial-based probes into food safety protocols not only enhances the precision of adulterant detection but also supports real-time monitoring, thereby ensuring compliance with stringent food safety regulations and protecting public health.

4.1.1 Metal–organic Frameworks (MOFs)

MOFs have emerged as a new class of porous coordinated polymer materials with large surface area and suitable for developing optical sensors. For example, Eu-TDA, a europium-based fluorescent MOF, is applied for the highly sensitive optical detection of malachite green (MG) and Uric acid (UA). The fluorescence emission of Eu-TDA was likely quenched by MG and UA due to spectral overlap, energy transfer, and competition between the species. The detection limit achieved for MG and UA is 0.0221 μM and 0.689 μM respectively [23]. In a different study, a rapid paper-based fluorescence sensor on terbium MOF-76 functionalized with aptamer for selective binding for the detection of MG was developed. Additionally, a MIP layer is formed around the MOF-76 and aptamer complex using MG as the template molecule providing dual recognition strategy. The detection mechanism is based on photo-induced electron transfer (PET). When MG bound to the aptamer and MIP recognition sites, the interaction triggered PET, leading to fluorescence quenching of the terbium MOF-76. The extent of fluorescence quenching directly correlated with the concentration of MG in the sample, enabling quantitative analysis. The limit of detection of present sensor was 0.1 ng/mL. Also, the method detects malachite green in fish and water samples with recoveries in 94.0–105.6% [17].

A label free and utltrasensitive fluorescent platform consisting of Tb@NH_2-MIL-253 (Al) MOF was fabricated to find the presence of melamine in milk samples. The sensor shows lower limit of detection 40 nM with in the time response of 30 s [2]. In another research work, Al-MOF and rhodamine (RhB) based fluorescent probe was synthesised and used in the development of smartphone based portable ratiometric fluorescent sensor. This sensor operates on a dual-emission mechanism, where the Al-MOFs and RhB serve as distinct fluorescent emitters. The Al-MOFs typically emit fluorescence at one wavelength (blue region), while RhB emits at another wavelength (orange region), which is attributed to ratiometric quenching induced by the inner filter effect (IFE) between Al-MOF/RhB and MG. The integration of this ratiometric sensor with a smartphone-based platform enhances its portability and user-friendliness. Smartphone cameras can capture the fluorescence emissions, and can analyse the emission ratios to quantify the presence of analytes. This approach facilitates on-site, real-time analysis without the need for complex laboratory equipment, making it particularly useful for applications in environmental monitoring, food safety, and clinical diagnostics. The fabricated sensor shows highly sensitive and selective detection in a wide range 0.5–200 μg/mL of MG with LOD 1.6 μg/mL. [103]. Similarly, blue to yellow color change dual-emission carbon dots and carboxymethyl chitosan encapsulated CDs have been used for the brilliant blue and sunset yellow detection [45, 106].

The incorporation of metal ions and other polymers into Metal–Organic Frameworks (MOFs) has significantly enhanced their optical properties, opening new avenues for their application in food adulterant detection. By introducing specific metal ions, the fluorescence characteristics of MOFs can be fine-tuned, improving their sensitivity and selectivity for various contaminants. For example, Eu^{3+} ions loaded on Cd-CP via post-modification to form Eu@Cd-CP, a coordination polymer-based fluorescent probe that can selectively detect sodium salicylate (SS) and sodium dehydroacetate (SA). These compounds commonly used in the processing of beverages, foods, and animal feeds to extend shelf life and inhibit certain bacteria. The probe can distinguish between SS and SA by quenching the fluorescence emission of Eu^{3+} ions via a photoinduced electron transfer process. The limits of detection for this Eu@Cd-CP probe are 1.25×10^{-6} M for SS and 8.33×10^{-7} M for SA [98]. Additionally, the integration of polymers with MOFs can provide a synergistic effect, enhancing the stability and functionality of the composite material. In a recent work, researchers have developed a polyethylenimine (PEI) modified MIL-101(Cr) composite based fluorescent sensor for determine meat adulteration. In the developed fluorescent sensor, pork-specific DNA sequence was used as the target, the surface potential of MIL-101 bio probe significantly affects its adsorption to 6-carboxyfluorescein-labeled DNA, indicating the presence of pork in camel meat, as shown in Fig. 3i. The detection mechanism involves a 6-carboxyfluorescein (FAM)-labeled DNA probe complementary to a pork-specific DNA sequence. In the absence of the target DNA, the FAM-labeled probe adsorbs onto the PEI-modified MIL-101(Cr) surface through electrostatic interactions, leading to fluorescence quenching due to the proximity of the fluorophore to the MOF surface. Upon introduction of the target pork DNA, hybridization occurs between the target and the FAM-labeled probe, forming a double-stranded DNA complex. This hybridization alters the surface potential and reduces the affinity between the double-stranded DNA and the PEI-MIL-101(Cr) composite, resulting in the desorption of the FAM-labeled probe from the MOF surface. Consequently, the fluorescence is restored, indicating the presence of pork DNA. The detection limit of 42 pM in just 90 min was achieved, without PCR amplification [40]. Moreover, a Cd based fluorescent MOF (QY = 74%) with strong emissive ligand 4,7-di(1,2,4-triazol-1-yl) benzo-2,1,3-thiadiazole was demonstrated for the detection of gossypol via PET luminescence quenching mechanism. The developed sensor displayed gossypol detection limit of 0.65 μM in the adulterated cottonseed oils [65].

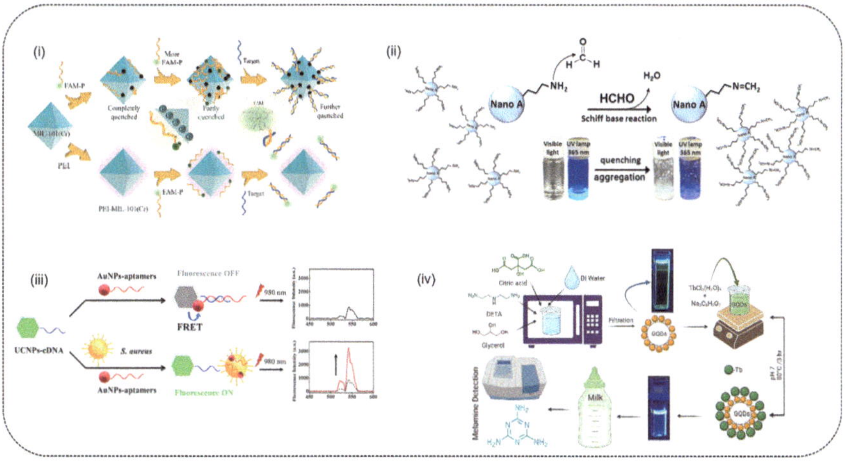

Fig. 3 Schematic diagram showing the fluorescence detection mechanism: **i** fluorescent PEI-MIL-101(Cr) based genetic detection of food adulteration; **ii** quantum dot-based sensitive formaldehyde detection in food samples; **iii** UCNPs-based sensitive detection of *S. aureus* via FRET; **iv** Tb-doped GQDs for the melamine determination in milk [5, 40, 43, 64]

4.1.2 Metal Nanoclusters

Comprising only a few to several hundred atoms, demonstrate distinctive properties that diverge from their bulk and larger nanoparticle counterparts. Owing to their diminutive size ($\leq$2 nm), these nanoclusters exhibit discrete electronic states and pronounced quantum confinement effects, which confer exceptional optical and electronic characteristics. Notably, these nanoclusters demonstrate high fluorescence, stability, and surface reactivity, resulting in highly suitable for applications in sensing, catalysis, and bioimaging. Also, their tunable properties and ease of functionalization further broaden their adaptability across diverse scientific domains. For instance, a ratiometric fluorescence sensor was developed used Copper nanoclusters (CuNCs) as a probe for detection of Curcumin and hypochlorite (ClO^-). The detection results show that addition curcumin in the quenches the fluorescence of CuNCs. The platform shows wide detection range of 0–412 μM with low detection limit of 24 μM. Furthermore, upon introducing ClO^- into the CuNCs-curcumin system, CuNCs' fluorescence was restored, and fluorescence of curcumin was quenched due to the oxidation of its phenolic and methoxy groups to quinones. The study also demonstrated a smartphone-assisted digital colorimetric analysis, which involved the acquisition, recognition, and processing of RGB data [47]. Other studies revealed that tiopronin-stabilized gold nanoclusters (TPN-AuNCs) and red-emitting Ag nanoclusters have been implied as fluorescent probe via quenching phenomena for the detection of melamine in infant formula [71, 97].

Additionally, a paper-based chemical sensing device with a fluorescent array has been developed to investigate various adulterants (CH_2O, H_2O_2 and NaClO,) in milk samples from different animal sources, including camel, sheep, goat, and cow. Electronic tongue systems, which utilize cross-selective sensing nanomaterials, have been designed to mimic the human sense of taste. This approach involves monitoring changes in the fluorescence of functionalized nanomaterials to detect milk adulteration from diverse species [21].

4.1.3 Quantum Dots (QDs)

Quantum dots (QDs) have also been widely used to develop several fluorescent sensors. In 2019 Xu et al. synthesised ZnCdSe-CdTe fluorescent QDs for rapid and sensitive detection of adulterants including syrups of glucose, sucrose, fructose, and glucose-fructose in honey samples. Their work established a benchmark by rapid and safely detecting even 1.0% of syrup adulterations in the honey samples [95]. Similar work by the same group has also studied the other fluorescent QDs based quenchers like CTAB-coated CdSe/ZnS QDs as fluorescence probes for adulterated vegetable oils. Their developed sensor with QDs as a probe shows LOD of 0.4% within 2 min of response time [96]. These researches demonstrate that use of semiconductor-based QDs (such as CdSe, CdTe, ZnSe, ZnS, etc.) can be very helpful for rapid and highly sensitive detection but suffer disadvantages like high toxicity, poor stability or biocompatibility, and severe photobleaching [11,

62, 94, 96]. Recently, carbon-based QDs have attracted considerable attention in the field of fluorescent nanosensors due to their biocompatibility, good photostability, and cost-effectiveness. CQDs also offer high fluorescence, excellent chemical stability, easy surface functionalization, and high quantum yields, enabling diverse uses in sensors, optoelectronics, and environmental monitoring applications [32, 44]. Furthermore, these QDs can be easily synthesized from variety of readily and inexpensive carbon-rich precursors such as organic molecules (e.g., citric acid, glucose) and natural sources (e.g., fruit peels, biomass). Their synthesis methods include low-cost hydrothermal/solvothermal synthesis, microwave-assisted synthesis, and chemical oxidation, which are relatively straightforward and often require mild conditions like moderate temperatures and ambient pressure [6, 16]. Several CQD-based sensors for fluorescence-based detection of food adulterants have been reported in recent years. For example, orange peel waste-derived CDs were synthesized to determine melamine. Ag^+ quenched the CDs' fluorescence due to the FRET energy transfer between CDs to Ag. The limit of detection of the method was ~0.85 μM and recovery percentage was found to be 98.03% [69]. The QDs are engineered by incorporating specific dopant elements into their semiconductor lattice to tailor their optical and electronic properties. These dopants, such as transition metals (e.g., Mn, Cu) or rare earth elements (e.g., Eu, Tb), introduce discrete energy levels within the bandgap of the QDs, significantly enhancing their luminescent characteristics. In addition, the doped QDs provide the free functional groups for the binding of biorecognition molecules and high phototstability ensures consistent detection over prolonged periods. For example, Tb-doped GQDs (size = 7 nm) were synthesized via simple conjugation microwave synthesis as illustrated in Fig. 3iv. This dual-emission fluorescent probe is able to detect a concentration of 0–5 μM of melamine in milk and milk products with a detection limit of 0.31 μM [5]. In a study, nitrogen (N)-doped carbon nanodots (N-CNDs) have been synthesized by hydrothermal treatment of prawn shells for application in NO_2^- in water. This proposed approach is based on dynamic fluorescence quenching, enabling a detection limit of 1.0 μM and a linear analytical range up to 1.0 mM. Similarly, yellow-emitting nitrogen-doped carbon dots (N-CDs, QY = 17.3%) and fluorescent MoS_2 QDs (QY = 20%) based selective detection of curcumin was developed where they could be quantitatively quenched by curcumin. This method detected curcumin in the range of 0.5–50 μM and 0.1–20 μM with a detection limit of 0.017 μM and 5 ng mL^{-1} [91, 110]. For example, Sulfur-doped CQDs were synthesized by using citric acid as the carbon source and thiourea as the sulfur precursor via a facile one-step microwave-assisted method. The prepared SCQDs were then used for the fabrication of a fluorescent probe; the sensor shows a detection limit of 0.77 M for Sudan I in saffron samples [51]. In another research established by Masoomi et al., a fluorescent array of CQDs with metal nanoclusters composites were explored as paper-based optical-nose for the quality control of saffron. These nanomaterials exhibit unique fluorescence properties that change when exposed to specific volatile organic compounds (VOCs) emitted by saffron. These fluorescence changes are detected and analyzed to create unique patterns corresponding to saffron quality and adulterants. Advanced data processing techniques are used to compare these patterns to reference profiles, enabling accurate assessment of saffron purity and identification of contaminants [49].

4.1.4 Upconversion Nanoparticles (UCNPs)

Upconversion nanoparticles (UCNPs) are nanoscale materials that convert low-energy photons, into higher-energy photons through a multiphoton absorption process. Their narrow emission bands and high chemical stability make them ideal for fluorescence applications. For example, a highly sensitive NaYF4: Yb active core–shell up-conversion nanoparticles immunochromatographic assay in the form of test strips was established to detect deoxynivalenol (a mycotoxin produced by *Fusarium* sp.) in cereals [77]. The strip-based platforms provide rapidness, simplicity and practical approach for their real-life applications. In addition, combining UCNPs with QDs can provide dual excitation, which can be utilized in the multiplex biosensing of bacterial pathogens. For instance, dual excitation-based UCNPs functionalized with specific aptamers and QDs were used to detect *S. typhimurium* and *S. aureus* [26, 59]. In a similar study, the S. aureus specific aptamer functionalized with AuNPs and was bonded to cDNA-modified UCNPs through complementary base pairing, resulting in fluorescence quenching. Upon introducing *S. aureus* into the system, the aptamers exhibited preferential binding to *S. aureus*, causing the dissociation of UCNPs-cDNA from the AuNP-aptamer complex and restoring fluorescence as illustrated in Fig. 3iii. This presented method showed a rapid and sensitive detection in range 47–4.7×10^7 CFU/mL with a detection limit 10.7 CFU/mL [5]. In another approach, different surface functional groups on UCNPs were used to identify pathogenic bacteria with varying phenotypes. Yin and coworker synthesised imidazole, phenylboronic and phosphate group functionalised UCNPs with distinct surface chemistries. The synthesised UCNPs is highly sensitive to the Gram-positive and Gram-negative bacteria such as *S. aureus* and *L. monocytogenes, E. coli, Salmonella*, and *V. parahaemolyticus* as foodborne pathogens. The interaction between the cell and the UCNPs increases the overall emission intensity of sensor signal thus indicating the presence of bacteria in the food. Each type of UCNP responded uniquely to different bacteria, generating specific luminescence signals [102].

4.1.5 Polymeric Nanomaterials

Polymeric nanomaterials have emerged as highly effective components in developing fluorescence-based optical probes for detecting food adulterants. These nanomaterials offer unique advantages, including high surface area, tunable optical properties, and the ability to interact specifically with various chemical compounds. For instance, polymers such as polystyrene, polyacrylate, and chitosan can be functionalized with fluorescent dyes or quantum dots to create sensitive and selective detection platforms. When these polymeric nanomaterials are exposed to food samples, they can emit fluorescence signals in response to the presence of specific adulterants, such as dyes, unauthorized additives etc. [24]. The fluorescence intensity or wavelength shifts provide a measurable indicator of contamination, enabling rapid and sensitive detection. This approach has been demonstrated in several studies, highlighting its potential for real-time monitoring and ensuring food safety. For example, fluorescent conjugated polymer dots were applied as the fluorescent probe for the detection of Ponceau 4R, a red synthetic food colorant in tomato sauces or pastes. The method could detect Ponceau 4R at very low concentrations, with a detection limit of 16 nmol/L [75]. In another research, a 3D fluorescence sensing method was exploited by developing a polymer probe consisting of hydrophilic hydrazine-naphthalimide functionalized chitosan (HN-chitosan) polymer and then used as onsite and rapid quantification of adulterants present in the saffron samples. Furthermore, a UV-cured polymeric membrane sensor composed of acrylate citric acid (ACA) units was synthesized specifically for detecting melamine in milk powder samples. This sensor demonstrated a response time of less than 1 min and exhibited a broad linear concentration range from 3.96×10^{-9} to 7.93×10^{-8} mol L^{-1}, with a detection limit of 2.32×10^{-1} mol L^{-1} [85]. In a related development, Jia and colleagues introduced a cellulose-based ratiometric fluorescent material for monitoring the freshness of seafood. This material exhibited a proportional, linear, reversible, and rapid response to various biogenic amines, such as ammonia, pyrrolidine, triethylamine, etc. Notably, NH_3 induced an immediate color change from red to orange or green, with the fluorescence intensity ratio of cellulose acetate immobilized with FITC and protoporphyrin IX linearly correlating with the logarithmic concentration of ammonia across a range of 5.0 ppm to 2.5×10^4 ppm. Furthermore, these polymer-based derivatives can be converted into diverse material forms, such as printable inks and membranes, into intelligent packaging and smart tags, enhancing their applicability in various fields [29]. Ye et al. discovered that fluorescent conjugated polymer nanomicelles could detect Sudan I dyes in water instead of natural pigments like capsanthin and beta-carotene, using Pluronic F127 as an encapsulating material. The fluorescence quenching occurs through a photochemical mechanism involving charge transfer between the polymer and dye molecules, which is further amplified by the hydrophobic core of the micelles [101].

4.2 SPR-Based Optical Probes

Surface plasmon resonance (SPR) sensors are non-destructive cutting-edge analytical devices that exploit the phenomenon of surface plasmons—coherent electron metal and a dielectric (non-conducting) medium interface. When polarized light strikes a thin metal film (typically gold or silver) at a specific angle, it induces surface plasmons, leading to a reduction in the reflected light intensity at that angle, known as the resonance angle. This reduction in the light intensity is proportional to the changes in the refractive index near the metal and a dielectric medium interface, making SPR sensors exceptionally effective for real-time monitoring and label-free detection of molecular interactions [34, 60]. The evanescent field generated during SPR extends only a few hundred nm into the surrounding medium, making it highly sensitive to molecular binding events occurring at or near the sensor surface as depicted in Fig. 4. When target molecules (analytes) bind to the sensor surface, which is often functionalized with specific ligands or receptors, the local refractive index changes, shifting the resonance angle. This shift can be detected and quantified, providing precise information about the presence and concentration of analytes [86].

In the past, prism-based SPR sensors (also known as Kretschmann configuration) have been employed to excite surface plasmons as presented in Fig. 4i. In an example, a SPR method is employed to detect 2–10% sugar adulterants (fructose, glucose, and sucrose) in pure honey. The experiment shows change in the resonance angle as a function of concentration of adulterants in honey samples. The results indicated high sensitivities 0.1266°/% for fructose, 0.1065°/% for glucose, and 0.0988°/% for sucrose [104]. In another study Singh and coworker presented a high refractive index based theoretical sensor with prism (ZnSe) as a dielectric substrate, Au particle nanolayer, Graphene thin film based monolayer and heterostructure of BlueP/MoS_2 for the determination of adulteration of milk [80]. Similarly, a detection system of prism/Au was developed for identification of adulterants in milk. The different concentrations of H_2O_2 were analysed with the sensitivity of 0.145°/mg ml^{-1} [76]. However, these traditional systems have several disadvantages, including their huge size due to bulky and large optical, electronic, and mechanical components and unsuitability in remote sensing areas [68].

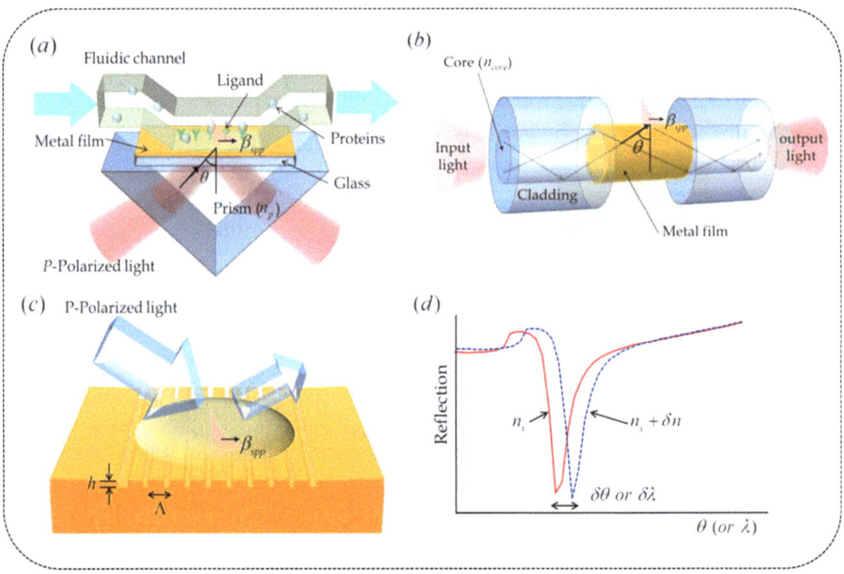

Fig. 4 Schematic illustration of SPR sensors representing **a** Kretschmann, **b** waveguide; **c** and grating-based coupling, respectively **d** graphical plot depicting light reflection ratio caused by SPR with angular or wavelength modulation [72]

Advancements in nanotechnology and integrated optics have developed more compact, robust, and efficient SPR sensors with improved designs and capabilities. Fiber-optic SPR sensors, in particular, offer enhanced portability, flexibility, and remote sensing abilities. Research has shown that using graphene as a dielectric overlayer can substantially increase the sensitivity of SPR sensor platforms, with enhancements of up to 3–4 orders of magnitude [88]. This approach not only stabilizes the metal in a liquid medium but also maintains the plasmonic resonances after the biofunctionalization of the sensor platform. Consequently, the deposition of graphene films over gold substrates has been shown to improve the performance of these sensors [18]. For instance, the detection of adulterants like glucose and fructose in honey samples can be made more sensitive by incorporating nanomaterials, such as Ag-graphene composites, into the sensor design [90]. Consequently, the employment of other nanomaterials, including two-dimensional (2D) materials and nanoparticles, has been explored further to increase the sensitivity of these SPR-based sensor systems. For example, graphene-$PtSe_2$ combined layers/ZnO nanosheet is presented in liquid solution. It was found that introducing the ZnO layer improved the overall light-collecting property, an appreciable shift in SPR angle, and improved sharpness of the SPR curve. The proposed sensor shows a sensitivity 155.33° RIU^{-1}, quality factor of 21.00 RIU^{-1}, and detection accuracy of ~0.62973 for formalin detection [54]. Salah and researchers developed an FO-SPR sensor incorporating Ag and molybdenum ditelluride ($MoTe_2$), which achieves a maximum sensitivity of 11,532.6 nm/RIU with the figure of merit of 80.647 RIU^{-1} for the detection of melamine in food products [25]. Other food contaminants like *Vibrio cholera*, ochratoxin A, and progesterone have also been successfully detected using graphene-based SPR sensors, proving its wide applicability in this field [14, 15, 38].

Almawgani et al. [3] presented an SPR-based sensor utilizing a Si/Ag dielectric layer and Ti_3C_2Tx MXene for detecting fat content in milk, achieving a maximum sensitivity of 350°/RIU for a 1.5% fat concentration. Another study developed a dual detection system for fat content and melamine in milk using transition-metal dichalcogenides (WS_2 and MoS_2) on an Ag-Pd bimetallic layer. This sensor platform achieved maximum sensitivities of 409.36°/RIU and 248.96°/RIU, with detection limits of 0.0085 RIU and 0.0140 RIU for fat and melamine, respectively [20]. An assay was developed on polycarboxybetaine acrylamide modified AuNPs attached with antibody for the detection of *E. coli* and *Salmonella* in food samples. The method shows the detection limit lowest to 57 CFU mL^{-1} and 17 CFU mL^{-1} *E. coli* and 7.4 $\times$ 10^3 CFU mL^{-1} and 11.7 $\times$ 10^3 CFU mL^{-1} for *Salmonella* sp. in cucumber and hamburger in 80 min [87].

Diffraction grating-based SPR sensors feature compact and integrated architectures with customizable wavelength control, whereas waveguide-based SPR configurations integrate seamlessly with microfluidic systems, providing enhanced sensitivity and scalability. For example, A highly sensitive FBG sensor, utilizing a Q-switched pulse erbium-doped fiber laser on spider silk, detects adulteration in antioxidant-rich kelulut honey by monitoring changes in the refractive index of isoglucose solution, causing a resonance shift in the FBG curve, which varies with optical wavelengths, affecting the resonance angle and half-width as shown in Fig. 4d

[55]. These advanced SPR sensor technologies collectively offer enhanced sensitivity, improved stability, and the ability for remote and in-situ monitoring applications. Increasing interest is being directed toward leveraging the characteristics and dimensions of optical fibers as highly sensitive sensors within generic microchannels [8, 33]. As a result, the integration of SPR with microfluidics and lab-on-a-chip technologies has revolutionized the optics field, enabling high-throughput, real-time analysis with minimal sample volumes. Mishra and the research group fabricated a microdevice featuring Z-shaped microfluidic channels integrated with optical fiber and microlens. This device exhibited a two-fold increase in sensitivity after immobilizing gold nanoparticles within the microfluidic channels. This microdevice demonstrated effective label-free detection of heavy metal ions (Pb^{2+}, Cr^{3+}, Hg^{+}) in water, with a minimum quantifiable concentration of 0.5 ppb [52]. Tang and colleagues developed a cost-effective integrated paper-lateral flow assay for nucleic acid testing detection. This device exhibited simple operation based on sample-in answer-out system on one paper to detect *S. typhimurium* in about an hour, with low detection limit [84].

In addition, integrating robust SPR sensing properties into smartphone platforms and internet-of-things (IoT) devices could yield an adaptive, ultrasensitive, and precise technology endowed with several advantages. These include remote, rapid, and facile detection, as well as the ability to onsite monitoring of environmental samples. Xiao et al. demonstrated the development of a smartphone-based surface Plasmon resonance imaging system that utilizes a 3D-printed monolithic microfluidic lab-on-a-chip platform integrated with a mobile phone to enable real-time monitoring of liquid samples for detecting food adulterants, environmental pollutants, and pathogens as depicted in Fig. 5iii [93]. Furthermore, for on-site detection of pork adulterants in the mixed meat, smart phone based integrated immunoassay was developed. This assay can detect very low levels of pork (0.01%) in meat mixtures within 30 min of the assay time. the developed immunoassay was able to detect as low as 0.01% of pork in meat mixtures in a total assay time of 30 min [74]. These advancements significantly broaden the scope and field deployable applications of SPR-based techniques in domains such as food safety and environmental monitoring.

4.3 SERS-Based Optical Probes

Surface-enhanced Raman spectroscopy (SERS) is considered one of the highly sensitive and selective optical techniques for detection at trace levels. Thus, a large number of researchers have explored SERS in various analytical chemistry applications, specifically identifying contaminants in food products [66]. SERS substrates are the important platform for amplifying the Raman signal of target molecules, and various materials have been developed for this purpose such as paper-based substrates (filter paper, chromatography paper majorly cellulose) are cost-effective, flexible, and disposable, making them ideal for on-site testing and rapid screening [89]. Nanomaterial coated thin film-based substrates, typically consists of thin layers of gold

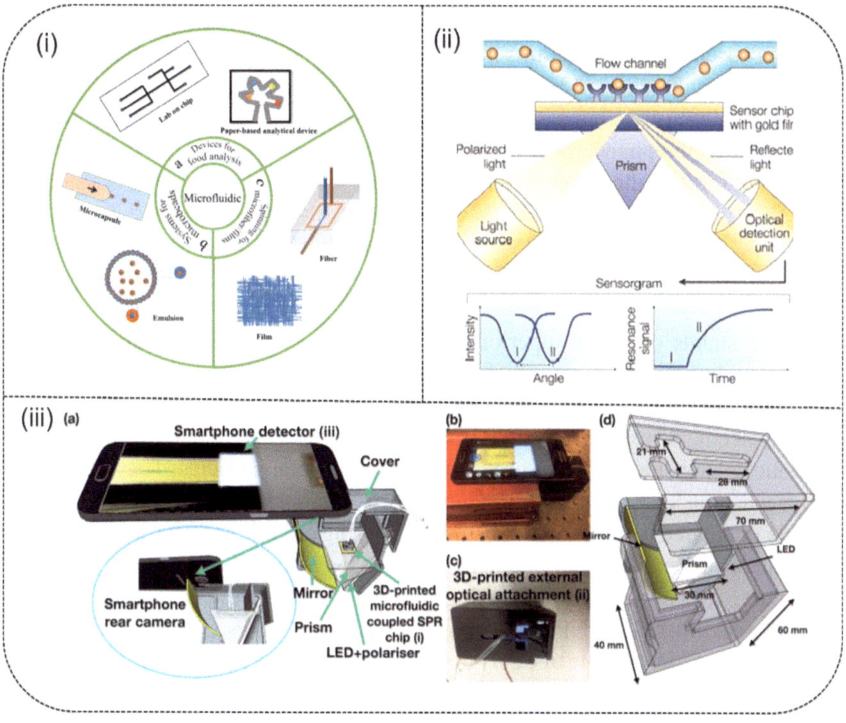

Fig. 5 **i** Applications of microfluidics as chips/ devices, films, optical fiber, microcapsules and emulsions in food safety; **ii** experimental setup of SPR (bio)sensor for microfluidic channels; **iii** illustration of a Smart iSPR system with 3D-printed SPR chip [13, 67, 93]

or silver nanoparticles on glass or silicon wafers, offer a uniform surface, ease of fabrication, and high reproducibility platform for used in laboratory for detailed analysis. These plasmonic nanoparticles exhibits localized surface plasmon resonance (LSPR), a phenomenon where conduction electrons collectively oscillate in response to incident light. This oscillation results in a concentrated enhancement of the electric field near the nanoparticle surface, significantly amplifying the Raman scattering signal when interacting with nearby molecules. The combination of the structural uniformity of thin-film substrates and the field enhancement through LSPR makes these platforms highly effective for precise detection [58]. For example, monolayer of Au NP–decorated filter paper with perfluorooctyltrichlorosilane modified fibers possess abundance of hotspot for high sensitivity. The fabricated platform shows enhancement factor of 1.63×10^8 with the detection limit in nanomolar concentration for rhodamine B in commercial beverages and paraquat herbicide in poisoned patients [10]. Further, inkjet-printed thin films of nanostructured silver particles act as flexible SERS substrates for the determination of food colorants Metanil Yellow and Malachite Green in food samples [35]. Moreover, Li and coworker developed a hemp-based spherical AgNPs SERS substrate that enables rapid detection of adulteration caused by food colorants in tea samples i.e. (carmine, sunset yellow, lemon yellow, and erythrosine) [42]. In addition, other advanced substrates include nanostructured materials like nanostars, nanoflowers, or nanocubes made of gold or silver, which offer high enhancement factors due to their well-defined nanostructures for identifying food adulteration [41, 50, 100, 113].

4.3.1 Graphene SERS Substrates

Graphene stands out as a superior SERS substrate due to its unique properties, including a large surface area, high electrical conductivity, and excellent chemical stability. It enhances Raman signals through two primary mechanisms: strong chemical enhancement involving charge transfer interactions with analytes and electromagnetic enhancement when paired with plasmonic nanoparticles. They also reduce fluorescence background, improve signal clarity, and provide better uniformity and stability compared to traditional substrates, ensuring reliable and reproducible results [39]. These properties make graphene a promising material for scalable, sensitive, and robust SERS applications in sensing domain. For example, a simple method using GO and popcorn-like AgNPs on a copper plate (GO/AgNPs/Cu) as a robust SERS detection substrate, capable of easily detecting trace residues of crystal violet (4.1 ng/g) and malachite green (3.6 ng/g) from contaminated fish scales [109]. Figure 6iii shows a grape skin-derived Au NP/GO-based paper SERS substrate was developed for onsite monitoring of rhodamine 6G in water and food contaminants (tap water, chilli powder, and tomato ketchup). The result shows 1.92×10^9, 7.33×10^{-11} M, and 10^{-11}–10^{-5} M enhancement factor, LOD, and linear range respectively [81]. Furthermore, Khalil et al. proposed a dual SERS biosensor based nanoplatforms consists of GO-gold nanorods functionalized capture probe and a uniquely designed Raman dye (ATTO Rho6G) conjugated with AuNPs. Their work indicates enhanced SERS intensity through both the electromagnetic hot spots generated at the junctions of the two platforms, with the chemical enhancement between the AuNPs and Raman tag. This platform demonstrated an impressive detection limit of 1 femtomolar when testing for DNA extracted from pork samples, enabling the ultra-sensitive detection of meat adulteration [31]. Hussain and co-workers simultaneously detected urea and ammonium sulfate in milk samples on AuNPs and bimetallic Au@AgNPs with a coffee ring effect, as shown in Fig. 6i. This effect occurs when a droplet is placed on a solid substrate and allowed to evaporate, forming concentric circles with solutes distributed along the radius. The capillary flow driven by evaporation moves solutes from the center to the edge, creating a self-assembled pattern of non-volatile materials. This process increases analyte concentration in the coffee ring region, enhancing the sensitivity of Raman techniques due to the combined effects of contact line pinning, solvent evaporation, and capillary flow [27].

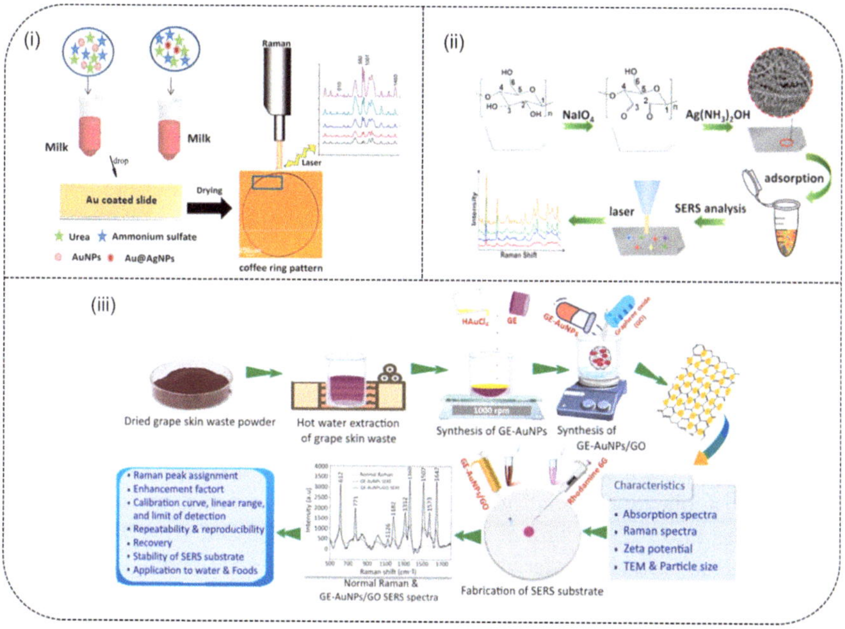

Fig. 6 **i** Detection of milk adulteration using coffee ring pattern-SERS with urea and ammonium sulfate; **ii** Diagram illustrating the fabrication process of a cellulose (paper)-based SERS substrate; **iii** Illustration showing the paper-based grape skin-AuNPs/GO SERS substrate for the determination of rhodamine dye [27, 63, 81]

4.3.2 Polymer SERS Substrate

Polymer-based substrates present promising opportunities for SERS-based detection of food adulteration. The use of polymers as templates for fabricating patterned nanostructures has been steadily increasing, driven by their ease of fabrication, flexibility, and tunable surface chemistry. Figure 6ii depicts the fabrication process of cellulose-based SERS substrate. Other polymers such as polystyrene, poly(methyl methacrylate) (PMMA), and poly(dimethylsiloxane) (PDMS) are commonly employed as SERS substrates due to these advantageous properties. These materials not only facilitate the formation of well-defined nanostructures but also enhance the sensitivity and adaptability of SERS platforms for diverse analytical applications. The key mechanism involves the encapsulation process, where hydrophobic interactions occur between a polymeric ligand shell and a compatible block copolymer. This interaction results in the formation of stable composite colloids in aqueous solutions, which are crucial for maintaining the structural integrity of the substrate. These substrates are designed to enhance the Raman signal of target analytes, providing high sensitivity and specificity [19, 48, 56]. For instance, polystyrene-based substrates coated with silver nanoparticles have been utilized to detect contaminants like melamine in milk (LOD 1.9×10^{-8} mol/L), demonstrating high sensitivity and rapid response times [111]. Similarly, PMMA substrates embedded with gold nanoparticles were used to detect pesticide residues in vegetables and fruits, leveraging the strong SERS enhancement of gold nanostructures [70, 105]. Moreover, PDMS substrates offer unique advantages due to their flexibility and transparency, making them suitable for both direct application onto food surfaces and integration with microfluidic devices. PDMS-based SERS substrates have been successfully used to detect adulterants in liquid samples, such as dyes in beverages and additives in dairy products [4, 37, 92], by simple swabbing of the sample surface or incorporating the substrate into a flow-through detection system.

While SERS is highly sensitive, robust, and suitable for sensitive onsite monitoring of food adulteration, it also has several limitations. The preparation of SERS-active substrates faces significant challenges due to the complex and costly processes involved, particularly the precise deposition of nanoparticles with standard approaches that often lack homogeneity, reproducibility, and hinders scale-up for the practical applications. Standard methods, such as the incorporation of plasmonic nanoparticles into the structure of flexible substrates or in situ growth of nanoparticles on the surface of bio-derived materials, do not provide homogeneity. Additionally, achieving consistent enhancement factors across different SERS substrates can be difficult, leading to variable results and challenges in standardization [66, 99]. Lastly, SERS signals are influenced by multiple factors, including substrate morphology, nanoparticle aggregation, and sample matrix composition, leading to highly complex spectra that are challenging to analyze using traditional chemometric approaches. Therefore, new fabrication techniques are still in demand. Future trends indicate the emergence of fields combining SERS with AI-driven machine learning algorithms, such as backpropagation neural networks, sparrow search algorithms, and convolutional neural networks, to enhance the determination of target substances [108].

5 Practical Applications

Portable Raman spectrometers have been increasingly employed in food authenticity testing across a wide variety of products. For instance, a portable 785 nm spectrometer successfully identified the adulteration of extra virgin olive oil with soybean oil [46]. In another study, a portable 785 nm wave-guided Raman spectrometer, in conjunction with a benchtop confocal micro-Raman system, was utilized for fish species authentication and melamine detection [28]. The integration of portable Raman spectrometers with chemometric analysis has also demonstrated significant potential in differentiating between authentic and counterfeit milk powder [30]. Notably, the Bravo handheld Raman spectrometer and the TruScan RM analyzer accurately classified genuine milk powder samples with success rates of 97% and 100%, respectively. In addition, portable Raman spectrometers have shown promise in assessing meat spoilage. A custom-designed miniature spectrometer operating at 671 nm was used to collect spectra from meat samples at different intervals up to three weeks post-slaughter. The study demonstrated successful discrimination between meat samples stored for varying durations [107]. These findings underscore the expanding utility of portable Raman spectroscopy in food quality and safety assessments. While Raman spectroscopy generates detailed biochemical fingerprints crucial for detecting food fraud, its sensitivity is inherently limited due to the low probability of Raman scattering events. Only about 1 in 10^6 to 10^8 photons undergo inelastic scattering, often necessitating signal enhancement techniques to improve detection capabilities. Despite these challenges, portable Raman spectroscopy represents a burgeoning field with significant potential in the food industry [82]. Fluorescence spectroscopy is another effective method for detecting food adulteration and ensuring food safety. For example, portable LED-based fluorescence spectroscopy has been used to measure the fluorescence intensity of pure and adulterated stingless bee honey samples, aiding in authentication [83]. Additionally, fluorescence spectroscopy has been employed to assess the quality of honey samples, specifically Sidr and unifloral (Acacia) honeys, and to detect sugar adulteration [1].

SERS is a highly effective vibrational spectroscopic method widely used for detecting food adulteration and ensuring food safety. Metrohm, a leading manufacturer of precision laboratory instruments, has been pivotal in advancing portable devices compatible with SERS technology. Among their notable innovations is the Metrohm Instant SERS Analyzer (MISA), which offers a straightforward, efficient, and accurate solution for rapid food authentication, delivering results within seconds [82].

6 Conclusion and Future Perspective

Food adulteration is a serious threat to consumers' health and safety, as it compromises the integrity and quality of food products. The nanomaterial-based optical probes like metal nanoparticles; 2D materials: Graphene-based materials/ MXenes/ Metal–organic frameworks/quantum dots and polymers based on fluorescence, SPR, and SERS provide desirable specificity with high sensitivity and rapid response time, which represent a transformative approach for the detection of food adulterants. Consequently, this can lead to the integration of in-situ detection probes with microfluidics and lab-on-a-chip technologies that have revolutionized the field, providing high-throughput, real-time analysis with minimal sample volumes. While these advanced sensors have been developed and studied extensively, their commercial availability varies. Some Raman and SERS-based sensors are available commercially for laboratory use, but widespread adoption in routine food adulteration detection is still emerging. Continuous advancements in sensor technology and integration with portable devices are expected to enhance their accessibility and application in food safety monitoring. Additionally, the last few years have also seen a growing trend in smartphone-based detection systems and IoT devices, which make these technologies accessible via user-friendly interfaces alongside data connectivity for remote monitoring and real-time analysis of food adulteration. SERS is an effective analytical technique for the detection of trace levels of contaminants and adulterants in food products with its impressive sensitivity and specificity enabled by numerous SERS substrates such as paper-based, nanomaterials films, polymer-substrate to graphene-substrate. Combining machine-learning algorithms with SERS shows promise in addressing these challenges, paving the way for more accurate and standardized analysis. The ongoing evolution of these nanomaterial-based optical probes, facilitated by advancements in nanomaterial design, integration with smart technologies, and the incorporation of artificial intelligence, holds substantial promise for ensuring food safety, safeguarding consumer health, and maintaining the integrity of food products.

References

1. H. Ali, K. Rafique, R. Ullah, M. Saleem, I. Ahmad, Classification of Sidr honey and detection of sugar adulteration using right angle fluorescence spectroscopy and chemometrics. Eur. Food Res. Technol. **248**(7), 1823–1829 (2022). https://doi.org/10.1007/S00217-022-04008-9/TABLES/1
2. M. Alizadeh Sani, G. Jahed-Khaniki, A. Ehsani, N. Shariatifar, M.H. Dehghani, M. Hashemi, H. Hosseini, M. Abdollahi, S. Hassani, Z. Bayrami, D.J. McClements, Metal–organic framework fluorescence sensors for rapid and accurate detection of melamine in milk powder. Biosensors **13**(1), 94 (2023). https://doi.org/10.3390/BIOS13010094
3. A.H.M. Almawgani, M.G. Daher, S.A. Taya, M. Mashagbeh, I. Colak, Optical detection of fat concentration in milk using MXene-based surface Plasmon resonance structure. Biosensors **12**(7), 535 (2022). https://doi.org/10.3390/BIOS12070535
4. A. Alyami, A.J. Quinn, D. Iacopino, Flexible and transparent surface enhanced Raman scattering (SERS)-Active Ag NPs/PDMS composites for in-situ detection of food contaminants. Talanta **201**, 58–64 (2019). https://doi.org/10.1016/J.TALANTA.2019.03.115
5. P.K.F. Anjila, G.R. Tharani, A. Sundaramoorthy, V. Kumar Shanmugam, K. Subramani, S. Chinnathambi, G.N. Pandian, V. Raghavan, A.N. Grace, S. Ganesan, M. Rajendiran, An ultra-sensitive detection of melamine in milk using rare-earth doped graphene quantum dots-synthesis and optical spectroscopic approach. Microchem. J. **196**, 109670 (2024). https://doi.org/10.1016/J.MICROC.2023.109670
6. S. Ashok Kumar, M. Dheeraj Kumar, M. Saikia, N. Renuga Devi, A. Subramania, A review on plant derived carbon quantum dots for bio-imaging. Mater. Adv. **4**(18), 3951–3966 (2023). https://doi.org/10.1039/D3MA00254C
7. S. Bansal, A. Singh, M. Mangal, A.K. Mangal, S. Kumar, Food adulteration: sources, health risks, and detection methods. Crit. Rev. Food Sci. Nutr. **57**(6), 1174–1189 (2017). https://doi.org/10.1080/10408398.2014.967834
8. M.A. Butt, N.L. Kazanskiy, S.N. Khonina, G.S. Voronkov, E.P. Grakhova, R.V. Kutluyarov, A review on photonic sensing technologies: status and outlook. Biosensors **13**(5) (2023). https://doi.org/10.3390/BIOS13050568
9. G. Campmajó, O. Núñez, *Chromatographic Fingerprinting Approaches in Food Authentication* (2021), pp. 137–165. https://doi.org/10.1142/9781786349958_0006
10. Y.H. Chen, C.C. Chen, L.C. Lu, C.Y. Lan, H.L. Chen, T.H. Yen, D. Wan, Wafer-scale fibrous SERS substrates allow label-free, portable detection of food adulteration and diagnosis of pesticide poisoning. Sens. Actuators, B Chem. **391**, 134035 (2023). https://doi.org/10.1016/J.SNB.2023.134035

11. M. Chern, J.C. Kays, S. Bhuckory, A.M. Dennis, Sensing with photoluminescent semiconductor quantum dots. Methods Appl. Fluorescence **7**(1), 012005 (2019). https://doi.org/10.1088/2050-6120/AAF6F8
12. K.T. Chung, Azo dyes and human health: a review. J. Environ. Sci. Health C **34**(4), 233–261 (2016). https://doi.org/10.1080/10590501.2016.1236602
13. M.A. Cooper, Optical biosensors in drug discovery. Nat. Rev. Drug Discov. **1**(7), 515–528 (2002). https://doi.org/10.1038/nrd838
14. D. Daems, J. Lu, F. Delport, N. Mariën, L. Orbie, B. Aernouts, I. Adriaens, T. Huybrechts, W. Saeys, D. Spasic, J. Lammertyn, Competitive inhibition assay for the detection of progesterone in dairy milk using a fiber optic SPR biosensor. Anal. Chim. Acta **950**, 1–6 (2017). https://doi.org/10.1016/J.ACA.2016.11.005
15. M.G. Daher, R. Manvani, S.K. Patel, H.B. Albargi, M. Jalalah, A.H.M. Almawgani, Super-sensitive detection of vibrio cholera using novel graphene-based optical device based on a surface Plasmon resonance structure. Plasmonics (2024). https://doi.org/10.1007/S11468-024-02204-9/METRICS
16. B.P. de Oliveira, F.O.M. da Silva Abreu, Carbon quantum dots synthesis from waste and by-products: perspectives and challenges. Mater. Lett. **282**, 128764 (2021). https://doi.org/10.1016/J.MATLET.2020.128764
17. N. Duan, X. Chen, X. Lin, D. Ying, Z. Wang, W. Yuan, S. Wu, Paper-based fluorometric sensing of malachite green using synergistic recognition of aptamer-molecularly imprinted polymers and luminescent metal–organic frameworks. Sens. Actuators, B Chem. **384**, 133665 (2023). https://doi.org/10.1016/J.SNB.2023.133665
18. Q. Duan, Y. Liu, S. Chang, H. Chen, J.H. Chen, Surface plasmonic sensors: sensing mechanism and recent applications. Sensors **21**(16), 5262 (2021). https://doi.org/10.3390/S21165262
19. S. Fateixa, H.I.S. Nogueira, T. Trindade, S. Fateixa, H.I.S. Nogueira, T. Trindade, SERS research applied to polymer based nanocomposites. Raman Spectroscopy (2018). https://doi.org/10.5772/INTECHOPEN.72680
20. L. Fatolahi, T.S. Addulrahman, S. Alemi, M.N. Al-Delfi, A.H. Athab, B.J. Janani, Optical detection of fat and adulterants concentration milk using TMDC (WS2 and MoS2)-surface Plasmon resonance sensor via high sensitivity and detection accuracy. Opt. Mater. **147**, 114723 (2024). https://doi.org/10.1016/J.OPTMAT.2023.114723
21. E. Ghohestani, J. Tashkhourian, B. Hemmateenejad, Rapid detection and quantification of milk adulterants using a nanoclusters-based fluorescent optical tongue. Food Chem. **456**, 139973 (2024). https://doi.org/10.1016/J.FOODCHEM.2024.139973
22. A. Haji, K. Desalegn, H. Hassen, Selected food items adulteration, their impacts on public health, and detection methods: a review. Food Sci. Nutr. **11**(12), 7534 (2023). https://doi.org/10.1002/FSN3.3732
23. L.J. Han, Y.J. Kong, G.Z. Hou, H.C. Chen, X.M. Zhang, H.G. Zheng, A Europium-based MOF fluorescent probe for efficiently detecting malachite green and uric acid. Inorg. Chem. **59**(10), 7181–7187 (2020). https://doi.org/10.1021/ACS.INORGCHEM.0C00620/ASSET/IMAGES/LARGE/IC0C00620_0006.JPEG
24. Himshweta, M. Singh, Nanosensor platforms for detection of milk adulterants. Sens. Actuators Rep. **5**, 100159 (2023). https://doi.org/10.1016/J.SNR.2023.100159
25. N. Hma Salah, Y. Vasimalla, B. Kaur, H.M. Rasul, C. Santhosh, R. Balaji, S. Kumar, HBL/NaF-based and molybdenum ditelluride-immobilized optical fiber SPR sensor for early detection of melamine residue in food products. Opt. Laser Technol. **179**, 111386 (2024). https://doi.org/10.1016/J.OPTLASTEC.2024.111386
26. J. Hu, F. Tang, Y.Z. Jiang, C. Liu, Rapid screening and quantitative detection of Salmonella using a quantum dot nanobead-based biosensor. Analyst **145**(6), 2184–2190 (2020). https://doi.org/10.1039/D0AN00035C
27. A. Hussain, D.W. Sun, H. Pu, SERS detection of urea and ammonium sulfate adulterants in milk with coffee ring effect. Food Addit. Contam. Part A **36**(6), 851–862 (2019). https://doi.org/10.1080/19440049.2019.1591643

28. Y. Hu, S.Y. Huang, X. Lu, A green analytical method for fish species authentication based on Raman spectroscopy. Green Anal. Chem. **4**, 100052 (2023). https://doi.org/10.1016/J.GREEAC.2023.100052
29. R. Jia, W. Tian, H. Bai, J. Zhang, S. Wang, J. Zhang, Amine-responsive cellulose-based ratiometric fluorescent materials for real-time and visual detection of shrimp and crab freshness. Nat. Commun. **10**(1), 1–8 (2019). https://doi.org/10.1038/s41467-019-08675-3
30. S.R. Karunathilaka, B.J. Yakes, K. He, L. Brückner, M.M. Mossoba, First use of handheld Raman spectroscopic devices and on-board chemometric analysis for the detection of milk powder adulteration. Food Control **92**, 137–146 (2018). https://doi.org/10.1016/J.FOODCONT.2018.04.046
31. I. Khalil, W.A. Yehye, N. Muhd Julkapli, A.A.I. Sina, S. Rahmati, W.J. Basirun, A. Seyfoddin, Dual platform based sandwich assay surface-enhanced Raman scattering DNA biosensor for the sensitive detection of food adulteration. Analyst **145**(4), 1414–1426 (2020). https://doi.org/10.1039/C9AN02106J
32. A. Khan, P. Ezati, J.T. Kim, J.W. Rhim, Biocompatible carbon quantum dots for intelligent sensing in food safety applications: opportunities and sustainability. Mater. Today Sustain. **21**, 100306 (2023). https://doi.org/10.1016/J.MTSUST.2022.100306
33. S.N. Khonina, N.L. Kazanskiy, M.A. Butt, Optical fibre-based sensors—an assessment of current innovations. Biosensors **13**(9) (2023). https://doi.org/10.3390/BIOS13090835
34. V.G. Kravets, A.V. Kabashin, W.L. Barnes, A.N. Grigorenko, Plasmonic surface lattice resonances: a review of properties and applications. Chem. Rev. **118**(12), 5912–5951 (2018). https://doi.org/10.1021/ACS.CHEMREV.8B00243/ASSET/IMAGES/LARGE/CR-2018-00243Y_0027.JPEG
35. A. Kumar, V. Santhanam, Paper swab based SERS detection of non-permitted colourants from dals and vegetables using a portable spectrometer. Anal. Chim. Acta **1090**, 106–113 (2019). https://doi.org/10.1016/J.ACA.2019.08.073
36. J.V. Kumar, J.W. Rhim, Fluorescent carbon quantum dots for food contaminants detection applications. J. Environ. Chem. Eng. **12**(2), 111999 (2024). https://doi.org/10.1016/J.JECE.2024.111999
37. S. Kumar, J.B. Kaushal, H.P. Lee, Sustainable sensing with paper microfluidics: applications in health, environment, and food safety. Biosensors **14**(6), 300 (2024). https://doi.org/10.3390/BIOS14060300
38. B. Lee, J.H. Park, J.Y. Byun, J.H. Kim, M.G. Kim, An optical fiber-based LSPR aptasensor for simple and rapid in-situ detection of ochratoxin A. Biosens. Bioelectron. **102**, 504–509 (2018). https://doi.org/10.1016/J.BIOS.2017.11.062
39. W. Liang, P. Wied, F. Carraro, C.J. Sumby, B. Nidetzky, C.K. Tsung, P. Falcaro, C.J. Doonan, Metal-organic framework-based enzyme biocomposites. Chem. Rev. **121**(3), 1077–1129 (2021). https://doi.org/10.1021/ACS.CHEMREV.0C01029/SUPPL_FILE/CR0C01029_SI_009.MP4
40. J. Li, Q. Sun, T. Jambal, D. Dorjgotov, N. Su, Q. Wu, Jirimutu, Rapid identification of adulterated camel meat by MIL-101(Cr)-based fluorescent sensing platform: surface potential adjustment to optimize detection performance. Microchem. J. **200**, 110476 (2024). https://doi.org/10.1016/J.MICROC.2024.110476
41. L. Li, W.S. Chin, Rapid and sensitive SERS detection of melamine in milk using Ag nanocube array substrate coupled with multivariate analysis. Food Chem. **357**, 129717 (2021). https://doi.org/10.1016/J.FOODCHEM.2021.129717
42. L. Li, Q. Cui, M. Li, T. Li, S. Cao, S. Dong, Y. Wang, Q. Dai, J. Ning, Rapid detection of multiple colorant adulteration in Keemun black tea based on hemp spherical AgNPs-SERS. Food Chem. **398**, 133841 (2023). https://doi.org/10.1016/J.FOODCHEM.2022.133841
43. Y.C. Lin, P.Y. Lin, S.L. Hsieh, C.Y. Tsai, A.K. Patel, R.R. Singhania, R. Kirankumar, C.D. Dong, C.W. Chen, S. Hsieh, Quantum dot assisted precise and sensitive fluorescence-based formaldehyde detection in food samples. Spectrochim. Acta Part A Mol. Biomol. Spectrosc. **283**, 121729 (2022). https://doi.org/10.1016/J.SAA.2022.121729

44. J. Liu, R. Li, B. Yang, Carbon dots: a new type of carbon-based nanomaterial with wide applications. ACS Cent. Sci. **6**(12), 2179–2195 (2020). https://doi.org/10.1021/ACSCENTSCI.0C01306/ASSET/IMAGES/LARGE/OC0C01306_0007.JPEG
45. L. Liu, M. Chen, T. Zhao, L. Yuan, Z. Mi, Y. Bai, P. Fei, Z. Liu, C. Li, L. Wang, F. Feng, Ratiometric fluorescence and smartphone-assisted sensing platform based on dual-emission carbon dots for brilliant blue detection. Spectrochim. Acta Part A Mol. Biomol. Spectrosc. **322**, 124782 (2024). https://doi.org/10.1016/J.SAA.2024.124782
46. Y. Li, T. Fang, S. Zhu, F. Huang, Z. Chen, Y. Wang, Detection of olive oil adulteration with waste cooking oil via Raman spectroscopy combined with iPLS and SiPLS. Spectrochim. Acta Part A Mol. Biomol. Spectrosc. **189**, 37–43 (2018). https://doi.org/10.1016/J.SAA.2017.06.049
47. Y. Li, Y. He, Y. Ge, G. Song, J. Zhou, Smartphone-assisted visual ratio-fluorescence detection of hypochlorite based on copper nanoclusters. Spectrochim. Acta Part A Mol. Biomol. Spectrosc. **255**, 119740 (2021). https://doi.org/10.1016/J.SAA.2021.119740
48. S. Mahanty, S. Majumder, R. Paul, R. Boroujerdi, E. Valsami-Jones, C. Laforsch, A review on nanomaterial-based SERS substrates for sustainable agriculture. Sci. Total Environ. **950**, 174252 (2024). https://doi.org/10.1016/J.SCITOTENV.2024.174252
49. S. Masoomi, H. Sharifi, B. Hemmateenejad, An optical-nose device based on fluorescent nanomaterials sensor array for authentication of saffron. Sens. Actuators, B Chem. **405**, 135365 (2024). https://doi.org/10.1016/J.SNB.2024.135365
50. M.M. Meenakshi, G. Annasamy, M. Sankaranarayanan, Highly sensitive technique for detection of adulterants in centella herbal samples using surface enhanced Raman spectroscopy (SERS). Spectrochim. Acta Part A Mol. Biomol. Spectrosc. **299**, 122878 (2023). https://doi.org/10.1016/J.SAA.2023.122878
51. T.G. Mir, A.Q. Malik, S. Shukla, J. Singh, D. Kumar, Facile synthesis of S-doped carbon quantum dots and their application in the detection of Sudan I in saffron. J. Fluoresc. **34**(1), 253–263 (2024). https://doi.org/10.1007/S10895-023-03264-6/METRICS
52. N. Mishra, A. Dhwaj, D. Verma, A. Prabhakar, Cost-effective microabsorbance detection based nanoparticle immobilized microfluidic system for potential investigation of diverse chemical contaminants present in drinking water. Anal. Chim. Acta **1205**, 339734 (2022). https://doi.org/10.1016/J.ACA.2022.339734
53. M. Momtaz, S.Y. Bubli, M.S. Khan, Mechanisms and health aspects of food adulteration: a comprehensive review. Foods **12**(1), 199 (2023). https://doi.org/10.3390/FOODS12010199
54. M. Moznuzzaman, M.R. Islam, I. Khan, Effect of layer thickness variation on sensitivity: an SPR based sensor for formalin detection. Sens. Bio-Sensing Res. **32**, 100419 (2021). https://doi.org/10.1016/J.SBSR.2021.100419
55. N.A.M. Muhammad, N.A. Awang, B.A. Talip, N.U.H.H. Zalkepali, M.I. Supaat, Highly sensitive adulteration detection in kelulut honey utilizing fiber Bragg grating technology with Q-switched pulse erbium-doped fiber laser and innovative spider silk substrate. Sens. Actuators, A **376**, 115656 (2024). https://doi.org/10.1016/J.SNA.2024.115656
56. R. Mu, N. Bu, J. Pang, L. Wang, Y. Zhang, Recent trends of microfluidics in food science and technology: fabrications and applications. Foods **11**(22), 3727 (2022). https://doi.org/10.3390/FOODS11223727
57. A. Nawrocka, J. Lamorska, A. Nawrocka, J. Lamorska, Determination of food quality by using spectroscopic methods. Adv. Agrophys. Res. (2013). https://doi.org/10.5772/52722
58. J. Neng, J. Wang, Y. Wang, Y. Zhang, P. Chen, Trace analysis of food by surface-enhanced Raman spectroscopy combined with molecular imprinting technology: principle, application, challenges, and prospects. Food Chem. **429**, 136883 (2023). https://doi.org/10.1016/J.FOODCHEM.2023.136883
59. L. Ngashangva, B.A. Hemdan, M.A. El-Liethy, V. Bachu, S.D. Minteer, P. Goswami, Emerging bioanalytical devices and platforms for rapid detection of pathogens in environmental samples. Micromachines **13**(7) (2022). https://doi.org/10.3390/MI13071083
60. H.H. Nguyen, J. Park, S. Kang, M. Kim, Surface Plasmon resonance: a versatile technique for biosensor applications. Sensors (Basel, Switzerland) **15**(5), 10481 (2015). https://doi.org/10.3390/S150510481

61. Q. Nie, S. Nie, High-performance liquid chromatography for food quality evaluation. Eval. Technol. Food Qual. (2019). https://doi.org/10.1016/B978-0-12-814217-2.00013-5
62. S.A. Nsibande, P.B.C. Forbes, Fluorescence detection of pesticides using quantum dot materials—a review. Anal. Chim. Acta **945**, 9–22 (2016). https://doi.org/10.1016/J.ACA.2016.10.002
63. S.A. Ogundare, W.E. van Zyl, A review of cellulose-based substrates for SERS: fundamentals, design principles, applications. Cellulose 26(11), 6489–6528 (2019). https://doi.org/10.1007/S10570-019-02580-0
64. Q. Ouyang, Y. Yang, S. Ali, L. Wang, H. Li, Q. Chen, Upconversion nanoparticles-based FRET system for sensitive detection of *Staphylococcus aureus*. Spectrochim. Acta Part A Mol. Biomol. Spectrosc. **255**, 119734 (2021). https://doi.org/10.1016/J.SAA.2021.119734
65. D.I. Pavlov, X. Yu, A.A. Ryadun, D.G. Samsonenko, P.V. Dorovatovskii, V.A. Lazarenko, N. Sun, Y. Sun, V.P. Fedin, A.S. Potapov, Multiresponsive luminescent metal–organic framework for cooking oil adulteration detection and gallium(III) sensing. Food Chem. **445**, 138747 (2024). https://doi.org/10.1016/J.FOODCHEM.2024.138747
66. A.I. Pérez-Jiménez, D. Lyu, Z. Lu, G. Liu, B. Ren, Surface-enhanced Raman spectroscopy: benefits, trade-offs and future developments. Chem. Sci. **11**(18), 4563–4577 (2020). https://doi.org/10.1039/D0SC00809E
67. N.M.M. Pires, T. Dong, U. Hanke, N. Hoivik, Recent developments in optical detection technologies in lab-on-a-chip devices for biosensing applications. Sensors **14**(8), 15458–15479 (2014). https://doi.org/10.3390/S140815458
68. B.A. Prabowo, A. Purwidyantri, K.C. Liu, Surface Plasmon resonance optical sensor: a review on light source technology. Biosensors **8**(3) (2018). https://doi.org/10.3390/BIOS8030080
69. A. Prasannan, T. Imae, One-pot synthesis of fluorescent carbon dots from orange waste peels. Ind. Eng. Chem. Res. **52**(44), 15673–15678 (2013). https://doi.org/10.1021/IE402421S/ASSET/IMAGES/LARGE/IE-2013-02421S_0007.JPEG
70. H. Qiu, M. Wang, S. Jiang, L. Zhang, Z. Yang, L. Li, J. Li, M. Cao, J. Huang, Reliable molecular trace-detection based on flexible SERS substrate of graphene/Ag-nanoflowers/PMMA. Sens. Actuators, B Chem. **249**, 439–450 (2017). https://doi.org/10.1016/J.SNB.2017.04.118
71. S.H. Ren, S.G. Liu, Y. Ling, N.B. Li, H.Q. Luo, Fluorescence detection of melamine based on inhibiting Cu^{2+}-induced disaggregation of red-emitting silver nanoclusters. Spectrochim. Acta Part A Mol. Biomol. Spectrosc. **201**, 112–118 (2018). https://doi.org/10.1016/J.SAA.2018.04.066
72. S. Roh, T. Chung, B. Lee, Overview of the characteristics of micro- and nano-structured surface Plasmon resonance sensors. Sensors **11**(2), 1565–1588 (2011). https://doi.org/10.3390/S110201565
73. S. Sargazi, I. Fatima, M. Hassan Kiani, V. Mohammadzadeh, R. Arshad, M. Bilal, A. Rahdar, A.M. Díez-Pascual, R. Behzadmehr, Fluorescent-based nanosensors for selective detection of a wide range of biological macromolecules: a comprehensive review. Int. J. Biol. Macromol. **206**, 115–147 (2022). https://doi.org/10.1016/J.IJBIOMAC.2022.02.137
74. N. Seddaoui, A. Amine, Smartphone-based competitive immunoassay for quantitative on-site detection of meat adulteration. Talanta **230**, 122346 (2021). https://doi.org/10.1016/J.TALANTA.2021.122346
75. V. Sedighi, F. Faridbod, Synthesis of polymer dots as fluorescent nanoprobe for the detection of Ponceau 4R, an additive color abuse in food. Food Chem. **434**, 137402 (2024). https://doi.org/10.1016/J.FOODCHEM.2023.137402
76. S. Sharma, A. Paliwal, M. Bassi, M. Tomar, V. Gupta, S. Gulati, Investigation of adulteration in milk using surface Plasmon resonance. ECS J. Solid State Sci. Technol. **10**(9), 091004 (2021). https://doi.org/10.1149/2162-8777/AC1F71
77. Y. Shen, X. Zhao, Z. Zhang, K. Fang, S. Chen, S. Tian, J. Fei, J. Zhu, A novel core-shell up-conversion nanoparticles immunochromatographic assay for the detection of deoxynivalenol in cereals. Talanta (2024). https://doi.org/10.1016/J.TALANTA.2024.125806
78. A. Shrivastava, *Frontiers in Drug Safety. Volume 1, Adulteration Analysis of Some Foods and Drugs* (2018), p. 197. https://books.google.com/books/about/Adulteration_Analysis_of_Some_Foods_and.html?id=JoJoDwAAQBAJ

79. N.A. Singh, N. Rai, A. Marwal, *Nanosensors for the Detection of Chemical Food Adulterants* (2021), pp. 25–53. https://doi.org/10.1007/978-3-030-69492-0_2
80. S.K. Singh, A. Srivastava, L.K. Dwivedi, A theoretical analysis of Milk adulteration/contamination detection in camel, buffalo and cow milk using SPR technique. J. Phys: Conf. Ser. **2426**(1), 012040 (2023). https://doi.org/10.1088/1742-6596/2426/1/012040
81. K. Sridhar, B.S. Inbaraj, B.H. Chen, An improved surface enhanced Raman spectroscopic method using a paper-based grape skin-gold nanoparticles/graphene oxide substrate for detection of rhodamine 6G in water and food. Chemosphere **301**, 134702 (2022). https://doi.org/10.1016/J.CHEMOSPHERE.2022.134702
82. J. Stradling, H. Muhamadali, R. Goodacre, Mobile guardians: Detection of food fraud with portable spectroscopy methods for enhanced food authenticity assurance. Vib. Spectrosc. **132**, 103673 (2024). https://doi.org/10.1016/J.VIBSPEC.2024.103673
83. D. Suhandy, D.F. Al Riza, M. Yulia, K. Kusumiyati, Non-targeted detection and quantification of food adulteration of high-quality stingless bee honey (SBH) via a portable LED-based fluorescence spectroscopy. Foods **12**(16), 3067 (2023). https://doi.org/10.3390/FOODS12163067
84. R. Tang, H. Yang, Y. Gong, M.L. You, Z. Liu, J.R. Choi, T. Wen, Z. Qu, Q. Mei, F. Xu, A fully disposable and integrated paper-based device for nucleic acid extraction, amplification and detection. Lab Chip **17**(7), 1270–1279 (2017). https://doi.org/10.1039/C6LC01586G
85. N. Taşci, S. Çubuk, E.K. Yetimoğlu, M.V. Kahraman, A novel polymeric fluorescence sensor based on acrylated citric acid for detection of melamine adulteration: application in milk powder. Food Chem. **394**, 133525 (2022). https://doi.org/10.1016/J.FOODCHEM.2022.133525
86. C.V. Topor, M. Puiu, C. Bala, Strategies for surface design in surface Plasmon resonance (SPR) sensing. Biosensors **13**(4) (2023). https://doi.org/10.3390/BIOS13040465
87. H. Vaisocherová-Lísalová, I. Víšová, M.L. Ermini, T. Špringer, X.C. Song, J. Mrázek, J. Lamačová, N. Scott Lynn, P. Šedivák, J. Homola, Low-fouling surface plasmon resonance biosensor for multi-step detection of foodborne bacterial pathogens in complex food samples. Biosens. Bioelectron. **80**, 84–90 (2016). https://doi.org/10.1016/J.BIOS.2016.01.040
88. A. Verma, B.C. Yadav, Comprehensive review on two dimensional nanomaterials for optical biosensors: present progress and outlook. Sustain. Mater. Technol. **40**, e00900 (2024). https://doi.org/10.1016/J.SUSMAT.2024.E00900
89. M. Verma, T.K. Naqvi, S.K. Tripathi, M.M. Kulkarni, P.K. Dwivedi, Paper based low-cost flexible SERS sensor for food adulterant detection. Environ. Technol. Innov. **24**, 102033 (2021). https://doi.org/10.1016/J.ETI.2021.102033
90. Vikas, M.K. Yadav, P. Kumar, R.K. Verma, Detection of adulteration in pure honey utilizing Ag-graphene oxide coated fiber optic SPR probes. Food Chem. **332**, 127346 (2020). https://doi.org/10.1016/J.FOODCHEM.2020.127346
91. S. Wang, X. Wang, K. Ning, G. Xiang, Fluorescent molybdenum disulfide quantum dots for sensitive detecting curcumin in food samples through FRET mechanism. J. Fluoresc. (2024). https://doi.org/10.1007/S10895-024-03720-X
92. H. Wu, C. Kanike, A. Marcati, X. Zhang, Flexible surface-enhanced raman scattering tape based on Ag nanostructured substrate for on-site analyte detection. Langmuir **40**(8), 4218–4227 (2024). https://doi.org/10.1021/ACS.LANGMUIR.3C03340/ASSET/IMAGES/LARGE/LA3C03340_0007.JPEG
93. C. Xiao, J. Eriksson, A. Suska, D. Filippini, W.C. Mak, Print-and-stick unibody microfluidics coupled surface plasmon resonance (SPR) chip for smartphone imaging SPR (Smart-iSRP). Anal. Chim. Acta **1201**, 339606 (2022). https://doi.org/10.1016/J.ACA.2022.339606
94. J. Xiong, H. Zhang, L. Qin, S. Zhang, J. Cao, H. Jiang, Magnetic fluorescent quantum dots nanocomposites in food contaminants analysis: current challenges and opportunities. Int. J. Mol. Sci. **23**(8), 4088 (2022). https://doi.org/10.3390/IJMS23084088
95. L. Xu, D. Lu, Q. Shi, H. Chen, S. Xie, G. Li, H.Y. Fu, Y.B. She, ZnCdSe-CdTe quantum dots: a "turn-off" fluorescent probe for the detection of multiple adulterants in an herbal honey. Spectrochim. Acta Part A Mol. Biomol. Spectrosc. **221**, 117212 (2019). https://doi.org/10.1016/J.SAA.2019.117212

96. L. Xu, X. Xu, H. Xiong, L. Chen, Y. Li, Rapid detection of vegetable cooking oils adulterated with inedible used oil using fluorescence quenching method with aqueous CTAB-coated quantum dots. Sens. Actuators, B Chem. **203**, 697–704 (2014). https://doi.org/10.1016/J.SNB.2014.07.008
97. X. Yang, Z. Jia, Z. Tan, H. Xu, N. Luo, X. Liao, Determination of melamine in infant formulas by fluorescence quenching based on the functionalized Au nanoclusters. Food Control **70**, 286–292 (2016). https://doi.org/10.1016/J.FOODCONT.2016.05.062
98. Y. Yang, D. Li, S. Qie, S. Su, M. Hu, Composite Eu@Cd-CP as a fluorescent probe for the detection of some food additives. Spectrochim. Acta Part A Mol. Biomol. Spectrosc. **317**, 124401 (2024). https://doi.org/10.1016/J.SAA.2024.124401
99. S. Yan, F. Chu, H. Zhang, Y. Yuan, Y. Huang, A. Liu, S. Wang, W. Li, S. Li, W. Wen, Rapid, one-step preparation of SERS substrate in microfluidic channel for detection of molecules and heavy metal ions. Spectrochim. Acta Part A Mol. Biomol. Spectrosc. **220**, 117113 (2019). https://doi.org/10.1016/J.SAA.2019.05.018
100. S. Yao-Say Solomon Adade, H. Lin, H. Jiang, S.A. Haruna, A. Osei Barimah, M. Zareef, A. Akomeah Agyekum, N. Adwoa Nkuma Johnson, M. Mehedi Hassan, H. Li, Q. Chen, Fraud detection in crude palm oil using SERS combined with chemometrics. Food Chem. **388**, 132973 (2022). https://doi.org/10.1016/J.FOODCHEM.2022.132973
101. X. Ye, J. Zhang, H. Chen, X. Wang, F. Huang, Fluorescent nanomicelles for selective detection of Sudan dye in pluronic F127 aqueous media. ACS Appl. Mater. Interfaces. **6**(7), 5113–5121 (2014). https://doi.org/10.1021/AM500270A/SUPPL_FILE/AM500270A_SI_001.PDF
102. M. Yin, C. Jing, H. Li, Q. Deng, S. Wang, Surface chemistry modified upconversion nanoparticles as fluorescent sensor array for discrimination of foodborne pathogenic bacteria. J. Nanobiotechnol. **18**(1), 1–14 (2020). https://doi.org/10.1186/S12951-020-00596-4/TABLES/1
103. X. Yue, Y. Li, S. Xu, J. Li, M. Li, L. Jiang, M. Jie, Y. Bai, A portable smartphone-assisted ratiometric fluorescence sensor for intelligent and visual detection of malachite green. Food Chem. **371**, 131164 (2022). https://doi.org/10.1016/J.FOODCHEM.2021.131164
104. N.H. Zainuddin, Y.W. Fen, A.A. Alwahib, M.H. Yaacob, N. Bidin, N.A.S. Omar, M.A. Mahdi, Detection of adulterated honey by surface plasmon resonance optical sensor. Optik **168**, 134–139 (2018). https://doi.org/10.1016/J.IJLEO.2018.04.048
105. D. Zang, C. Antonini, A.M. Coclite, M. Matamoros-Ambrocio, E. Sánchez-Mora, E. Gómez-Barojas, Surface-enhanced Raman scattering (SERS) substrates based on Ag-nanoparticles and Ag-nanoparticles/poly (methyl methacrylate) composites. Polymers **15**(12), 2624 (2023). https://doi.org/10.3390/POLYM15122624
106. J. Zhang, H. Qi, T. Yi, T. Jing, M. Zhao, J. Li, M. Ran, X. Zhu, C. Luo, Development of a highly sensitive ultra-small ratiometric fluorescence nanosphere probe for Sunset Yellow detection. Talanta **277**, 126341 (2024). https://doi.org/10.1016/J.TALANTA.2024.126341
107. L. Zhang, Q. Yu, M. Zhang, C.L. Law, Y. Ma, Intelligent detection of quality deterioration and adulteration of fresh meat products in the supply chain: research progress and application. Food Biosci. **55**, 103047 (2023). https://doi.org/10.1016/J.FBIO.2023.103047
108. L. Zhang, C. Zhang, W. Li, L. Li, P. Zhang, C. Zhu, Y. Ding, H. Sun, Rapid identification of Auramine O dyeing adulteration in *Dendrobium officinale*, saffron and curcuma by SERS Raman spectroscopy combined with SSA-BP neural networks model. Foods **12**(22), 4124 (2023). https://doi.org/10.3390/FOODS12224124/S1
109. M. Zhang, Z. Chen, Z. Wang, Z. Zheng, D. Wang, Graphene oxide coated popcorn-like Ag nanoparticles for reliable sensitive surface-enhanced Raman scattering detection of drug residues. J. Mater. Res. **34**(17), 2935–2943 (2019). https://doi.org/10.1557/JMR.2019.78/METRICS
110. S. Zhang, J. Bao, L. Lv, Y. Bai, Y. Zhang, Y. Zhang, Yellow-emitting carbon dots as fluorescent sensors for the rapid determination of curcumin. Spectrochim. Acta Part A Mol. Biomol. Spectrosc. **309**, 123819 (2024). https://doi.org/10.1016/J.SAA.2023.123819
111. Y. Zhao, W. Luo, P. Kanda, H. Cheng, Y. Chen, S. Wang, S. Huan, Silver deposited polystyrene (PS) microspheres for surface-enhanced Raman spectroscopic-encoding and rapid label-free

detection of melamine in milk powder. Talanta **113**, 7–13 (2013). https://doi.org/10.1016/J.TALANTA.2013.03.075

112. J. Zhou, Y. Gui, X. Lv, J. He, F. Xie, J. Li, J. Cai, Nanomaterial-based fluorescent biosensor for food safety analysis. Biosensors **12**(12) (2022). https://doi.org/10.3390/BIOS12121072
113. N. Zhou, H. Zheng, M. Hu, R. Liu, G. Qin, Y. Lu, H. Tang, Z. Huang, Ag nanopines over Al nanoneedle arrays enable high SERS sensitivity and reproducibility for molecular sensing. Surf. Interfaces **49**, 104457 (2024). https://doi.org/10.1016/J.SURFIN.2024.104457

Bridging Technology and Food Safety: An Overview of Smartphone-Based Fluorescence Sensing for Allergen Detection

Sibasish Dutta

Abstract Smartphone-based fluorescence sensing is a promising technology for the detection of food allergens. This approach leverages the ubiquity and advanced capabilities of smartphones, combined with novel sensing technologies, to provide a portable, low-cost, and user-friendly platform for allergen detection. The technology involves the use of biosensors and fluorescence signal-enhancing substrates integrated with a smartphone. The smartphone's camera and embedded flash functions are used to capture images and quantify the fluorescence response, enabling real-time, on-site analysis. This system can identify allergens by their unique fluorescence signatures when exposed to certain wavelengths of light. One of the key advantages of this technology is its potential for widespread use. The simplicity and affordability of the system make it accessible to a broad range of users, from industry professionals to consumers. This could revolutionize food safety, allowing individuals to test their food for allergens wherever and whenever they need. However, while the technology is promising, further research and development are needed to improve its sensitivity and specificity, and to expand the range of allergens it can detect. This chapter summarizes the current trends of smartphone-based fluorescence techniques for foods allergens detection, its limitations and future scopes.

Keywords Smartphone · Fluorescence · Allergens · Camera

1 Introduction

Exposing to a specific allergen due to ingesting certain food can trigger type I hypersensitivity which is a significant health concern. Such allergic reactions are induced due to the production of IgE antibodies which bind to allergens. As a result, immune cells such as mast cells and basophils are activated causing release of histamines

S. Dutta (✉)
Department of Physics, Pandit Deendayal Upadhyaya Adarsha Mahavidyalaya (PDUAM)
Eraligool, Sribhumi 788723, Assam, India
e-mail: dutta.sibasish94@gmail.com

R. Biswas (ed.), *Optical Techniques for Assessing Food Adulterants*, Progress in Optical Science and Photonics 35, https://doi.org/10.1007/978-981-96-8142-6_7

and other chemicals manifesting into allergic symptoms. It has been observed that there is a difference between allergic reactions caused due to different types of food amongst children and adult. While children are susceptible to wheat, soy, milk and eggs, adults may react to nuts (almonds and walnuts), buckwheat, sesame seeds and other different types of foods. Hence, it can be concluded that there is a huge diversity of allergic proteins.

The complexity of food allergens can be estimated by observing it's variation among different class of individual and population. What may be safe for some can trigger severe reactions in others. As limited studies have been made on allergic potential of protein, there can be significant variations among population reacting to such proteins that may require extensive study. Even the allergome database prepared based on IgE binding test compiling a list of allergens is not sufficient enough to account for allergenicity across different sections of population. Therefore, a need of comprehensive studies covering more larger and diverse group of population for understanding and predicting the allergic response is required. Considering global health risks worldwide, current research and technological innovations are critical for leveraging our understanding and hence management of these conditions.

To mitigate these challenges, technological advancement offers a promising solution. Some of the most commonly established gold standard analytical techniques for food allergen detection are ELISA [1], LC–MS [2] and PCR [3]. However, researchers around the world have further explored graphene oxide-based sensors [4–6], SPR sensors [7], aptamer based [8–10], and microfluidic devices [11] for detection and estimation of food allergens. Although the established analytical techniques and the currently emerging techniques are proven to be efficient enough to detect food allergens, such techniques lack user-friendliness, simplicity and data transmission facility combined with ease of operation which is vital for detection of allergens in the emerging world of point-of-care diagnosis. Towards this end, smartphone offers a promising solution to cater the need of masses and health care providers for enhanced diagnosis and treatment strategies through navigating the complex world of food allergens. Additionally, smartphone-based allergen devices offer user-friendly interface with simplicity and data transmission facilities through its cellular network. Although a number of articles have been reported on food allergen detection using smartphones [1, 12–17] no articles have been reported specifically dedicated to smartphone-based fluorometric detection of food allergens. In this book chapter, a focused review has been done on fluorescence sensing of food allergens using smartphone as optical reader.

2 Smartphone as Fluorescence Signal Reader for Food Allergen Detection

The binding of food allergens with fluorescent probes made of quantum dots (QDs) can trigger change in the fluorescent intensity in form of quenching. With increase in analyte concentration, decrease in fluorescence intensity occurs. Thus, the degree of fluorescence quenching can be inversely correlated with allergen concentration. Further, since the fluorescent signals fall in visible domain (400–700 nm) of electromagnetic spectrum, the camera of smartphones can easily detect these variations. The smartphone camera has IR filter which limits the incident wavelength in the range 400–700 nm only. Photodiodes used in the image sensor of smartphone records light intensity in gray scale values in terms of 8-bit (0–255 levels). The gray scale value is a measure of intensity of light. Darker shades represent lower intensity while lighter shades represent higher intensity of light. It is possible to determine the extent of fluorescence quenching by analyzing the grey scale values. Thus, decrease in light intensity due to fluorescence quenching of the target analyte can be captured by smartphone camera which can be correlated with concentration.

With readily available color detection apps from Google Playstore & Apple App store or by developing custom designed app, those signals or image captured can be analyzed through different color channels namely RGB, HSV, HSL, CMYk etc. Smartphone's camera is designed to mimic the function of human eye with sensitivity across shorter, middle and longer wavelength. The design is based on placing a Bayer color filter on the CMOS image sensor which can one of the three primary colours. Figure 1 shows schematic representation of smartphone-based fluorescence sensing of food allergens.

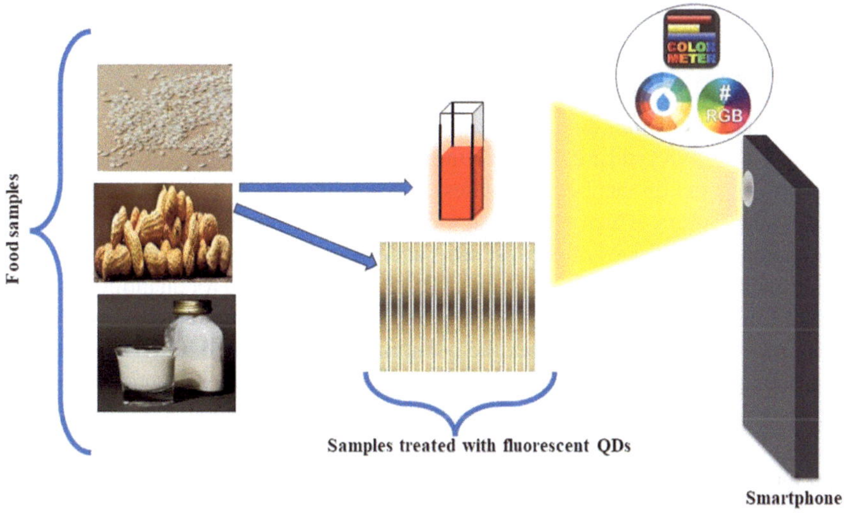

Fig. 1 Schematic representation of smartphone-based fluorescence sensing of food allergens

3 State-of-the-Art on Smartphone-Based Fluorescence Sensing of Food Allergens

Quantum Dots (QDs) or Carbon Quantum Dots (CQDs) are the most commonly used fluorescent probes for food allergens detection applications [9, 18, 19]. Selective binding of functionalized quantum dots with specific allergic proteins exhibits change in fluorescence properties that can be measured and quantified with smartphone cameras. The unique optical properties of quantum dots offer high sensitivity and specificity for detection of food allergens. Integrating quantum dots with microfluidic systems and lateral flow assay can leverage the detection capabilities of quantum dots [20–22]. Similar techniques could be utilized for detection of other food contaminates. In such techniques food contaminants getting specifically binded to fluorescent probes can be detected through ratiometric fluorescent signal detection [23] and on–off-on mode [24].

The change in fluorescent intensity with allergen concentration can be easily detected with camera of smartphones and analyzed through custom designed apps or even with some freely available apps. Although sensitivity of such methods is comparable to that of established techniques, there are challenges and considerations for practical applications in food safety and allergen management. In the subsequent part of this chapter, different methods reported for fluorometric detection of food allergens on smartphone platform has been discussed.

3.1 *Test Strip*

The first ever sesame allergen detection on a smartphone platform through fluorescence was reported by Linghu et al. [25]. They developed a test strip for sensitive detection of sesame allergen in food samples. The test strip is composed of a sample pad, conjugate pad, nitrocellulose (NC) membrane, and absorbent pad, all affixed to a PVC sheet. Quantum dots (QDs) labeled with ovalbumin (OVA) form the signal probes (QDs-OVA) and are coated on the C-line. A mixture of QDs-OVA and sesame protein is coated on the T-line. The sesame protein in the sample causes change in fluorescence signal which is a measure of analyte concentration. The smartphone's camera captures images of the test strip under UV light, and an app calculates the color intensity in RGB values. The mean RGB intensity is used for quantification. The visual LoD for sesame protein was 80 μg/L, while the quantitative LoD was 40 μg/L. In real food sample detection, the quantitative LoDs were 320 μg/L for biscuits and 640 μg/L for bread and ham. With high specificity and minimal cross-reactivity, the method showed repeatability between 80 and 120% as compared to conventional ELISA kit. However, the test requires careful optimization of parameters such as pH and the amount of AuNPs-Ab. It also necessitates professional technical personnel for accurate interpretation of results. Figure 2 shows the fluorometric detection of sesame allergen through test strips detected using smartphone.

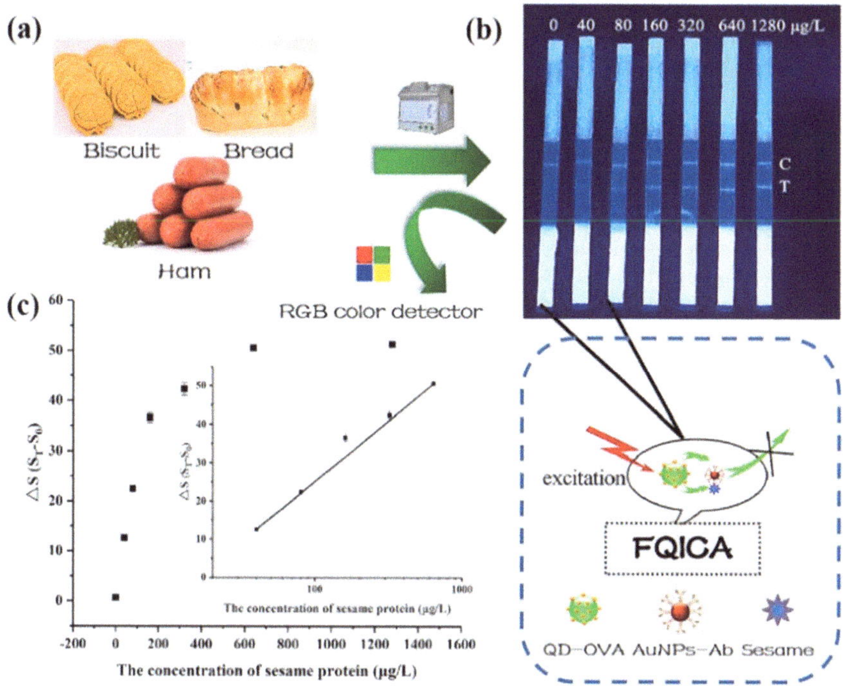

Fig. 2 Fluorometric detection of Sesame allergen through test strips. Reproduced from Ref. [25] with permission from Elsevier

3.2 Assay with Magnetic Separation

Linghu et al. [26] reported another method of detection of sesame allergen by using ZnCdSe/ZnS quantum dots as signal markers. In this method, two types of probes were used namely, signal probes and capture probes. The synthesis of capture probes involved modifying carboxyl-functional magnetic polystyrene microspheres with sesame allergens. These microspheres allowed separation of target allergens from the sample. On the other hand, ZnCdSe/ZnS quantum dots (QDs) covalently bonded with sesame allergen monoclonal antibodies served as signal probes. Upon interaction with allergens, these QDs emit fluorescence signals whose intensity changed with change in concentration of sesame protein. A readily available Color Picker APP was used to quantify the fluorescence intensities. With LoD of 10.15 μg/L and linearity in the range 80–640 μg/L, the assay recovery rates were found to be in the range 89–109%, indicating high repeatability and consistency. However, additional dilutions or additional processing steps may be required to ensure sensor's accuracy against high concentration of matrix ions in food samples. Figure 3 shows the fluorometric detection of sesame allergen using ZnCdSe/ZnS quantum dots as signal markers in combination with smartphone as optical reader.

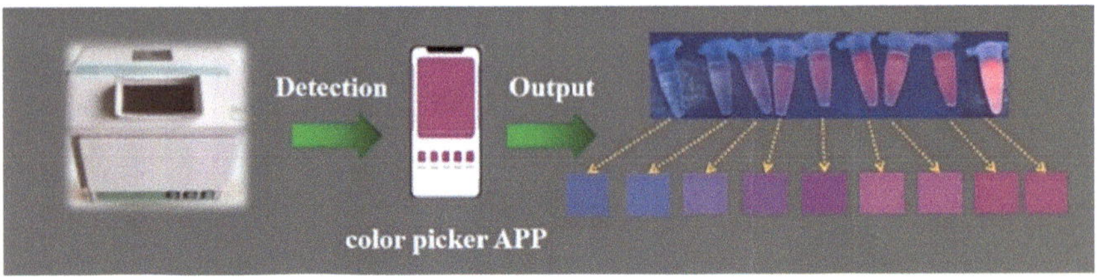

Fig. 3 Assay with magnetic separation for detection of QDs ZnCdSe/ZnS quantum dots as signal markers. Reproduced from Ref. [26] with permission from Elsevier

3.3 Paper Sensor Chip

Detection of β-LG through fluorescence intensity on a smartphone platform was reported by Yang et al. [27]. Here, QD-Apt fluorescent probe was mixed with β-LG solution, followed by addition of graphene oxide (GO) solution. The mixture was dispensed on to a sensor chip followed by drying, allowing the formation of visible coffee ring. The function of the sensor was based on fluorescence quenching where intensity of the fluorescence signal decreases with increasing concentration of analyte.

A smartphone camera was used for capturing the image of the coffee ring and analyzed for correlation with allergens. The LoD for β-LG was determined to be 0.048 mg L.

The assay demonstrated a linear range from 0.39 to 1000 mg L^{-1}, indicating good sensitivity.

The method showed consistent results with those of a commercial ELISA assay kit, suggesting good repeatability. However, such sensors may face challenges like interference from other substances in food samples or the need for precise environmental control during measurements. Figure 4 shows the fluorometric detection of β-LG using QD-Apt as fluorescent probe in combination with smartphone as detection system.

3.4 μPAD

Gomez et al. [28] reported β-LG detection by developing an aptasensor using a microfluidic paper-based analytical device (μPAD) with a β-lactoglobulin (β-LG) specific aptamer immobilized on a cellulose disk. Fluorescent Carbon Quantum Dots (CQDs) bioconjugated with complementary DNA were used as a label for aptamer hybridization. A smartphone camera or digital camera captures the fluorescence intensity changes for β-LG concentration estimation.

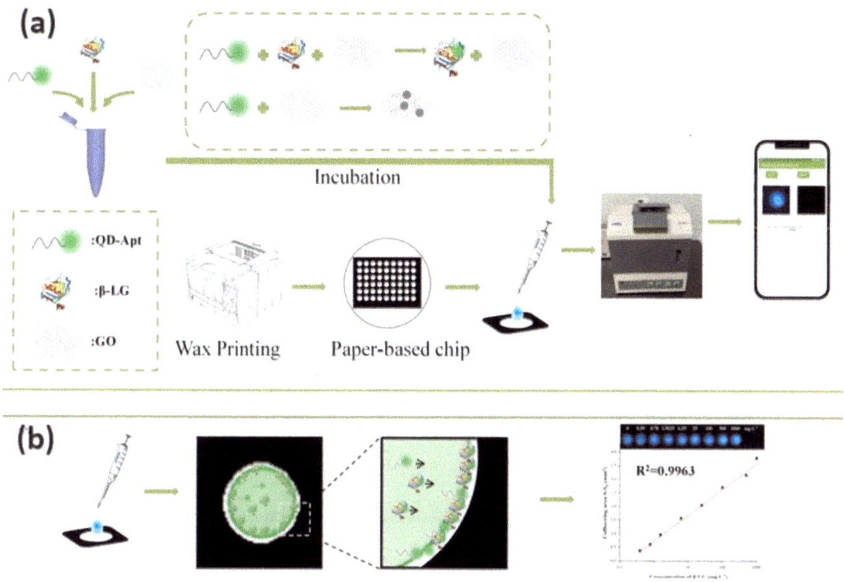

Fig. 4 Detection of β-LG using QD-Apt fluorescent probe. Reproduced from Ref. [27] with permission from Elsevier

The sensor gave an LoD of up to 0.6 ng/mL and exhibited excellent detection response within the range of 1.8 to 500 ng/mL. However, a decrease in performance of the sensor after 28 days indicated the limitation in its stability. Figure 5 shows the fluorometric detection of β-LG in μPAD platform with smartphone as detection system (Table 1).

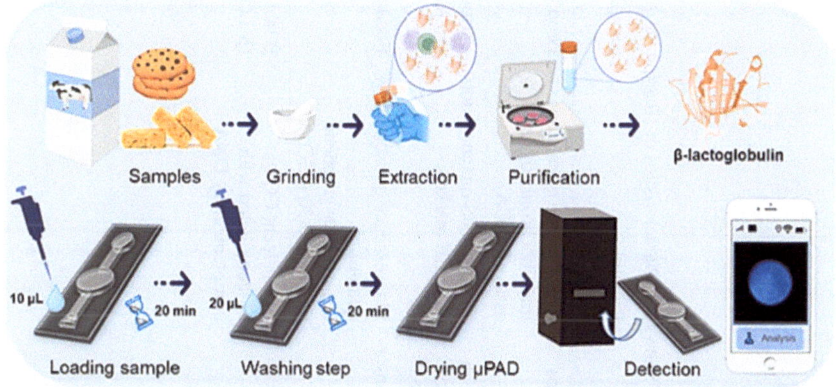

Fig. 5 μPAD platform for detection of β-LG. Reproduced from Ref. [28] with permission from Elsevier

Table 1 Shows the characteristics of smartphone-based allergen detection through fluorescence as reported by different researchers around the world till date

Device	Principle	Sample	Target	Linear Range	Limit of detection	Repeatability	References
Test strips	Fluorescence quenching immunochromatographic assay (FQICA)	Food	Sesame allergens	80–640 μg/L	40 μg/L	80–120% recovery	[25]
Assay with magnetic separation	ZnCdSe/ZnS quantum dots as signal markers	Various food samples	Sesame allergens	80–640 μg/L	10.15 μg/L	89–108% recovery	[26]
Paper sensor chip	Fluorescence quenching by quantum dot-aptamer-graphene oxide (QD-Apt-GO)	Foods	Beta-lactoglobulin (β-LG)	0.39–1000 mg/L	0.048 mg/L	–	[27]
μPAD with β-LG aptamer	Fluorescent CQDs bioconjugated with complementary DNA	β-lactoglobulin (β-LG)	β-lactoglobulin concentration	1.8–500 ng/mL	0.6 ng/mL	RSD: 5.3% (repeatability), 8.3% (reproducibility)	[28]

4 Conclusion

Fluorescence sensing of food allergens has been a popular technique for detecting variety of analytes (allergens) in food samples. The incorporation of smartphone as optical reader can truly enhance the potential for on-site detection and estimation of food allergens. Till date, two different types of allergens namely sesame and β-LG were estimated by correlating the intensity of fluorescence signal with allergen concentration. Sesame allergen detection using smartphone through fluorescence quenching immunochromatographic assay (FQICA) on a test strip was found to be effective in detecting sesame allergens with high sensitivity and specificity. On the other hand, preparation of assay using ZnCdSe/ZnS quantum dots as signal markers was found to be a rapid, sensitive, and portable method for sesame allergen detection, suitable for both laboratory and field testing. Further paper sensor chip based β-LG detection through fluorescence quenching of quantum dot-aptamer-graphene oxide (QD-Apt-GO) provided a simple, economic, and visual detection platform for β-LG in foods, suitable for use in resource-limited settings. Also, μPAD with β-LG aptamer was an innovative aptasensor providing a portable, low-cost, and easy-to-use solution for on-site monitoring of β-LG in food products, meeting the demands of food safety analysis. As a screening tool, camera of the smartphone with specific app installed for signal readout enhanced its practicality for on-site testing. Such devices, can be very useful in resource limited settings to ensure meeting the demands of food safety analysis.

5 Limitations and Future Scopes

Smartphone based allergen detection techniques reported till date are although opening new avenues for on-site allergen analysis in food samples but still, it has a long way to go because of the following limitations. Firstly, such techniques as a whole may have to deal with optimization challenges as the test requires careful optimization of parameters such as pH and the amount of QDs-Ab (antibody-conjugated Quantum Dots). Achieving consistent and reliable results across different food samples and conditions can be challenging due to variations in these parameters.

Secondly, interference from matrix ions is another issue to deal with. High concentrations of matrix ions in food samples can interfere with the sensor's accuracy. These ions may lead to false positives or negatives, necessitating sample dilution or additional processing steps to mitigate their impact. Thirdly, precise environmental control during measurements is crucial for accurate results. Factors like temperature and humidity can affect the sensor's performance. Smartphone-based systems may struggle to maintain consistent conditions compared to dedicated laboratory equipment. Fourthly, the sensor's performance may decrease after a certain interval of time. This suggests a limitation in long-term stability. Users should consider the shelf life and reliability of the samples over extended periods. The consistency and reliability of the results can also be influenced by users' variability. As variability in technical expertise in the form of sample handling, precisely following testing protocols and interpreting the results can also lead to variations in output. Cost implications should also be considered especially during initial manufacturing phases of detection kits involving fluorescence enhancing substrates. It is imperative to ensure that detection kits are affordable for consumers.

Environmental conditions can be mitigated by developing robust calibration algorithms capable of adjusting variations due to temperature, humidity and ambient light. To reduce the influence of users' variability, comprehensive, user-friendly and standardized procedures can be developed, guiding the user through flowcharts and visual aids. Further, detection kits subsidization can be realized through partnership with industries and various Govt. agencies.

Leaving aside the limitations introduced in the process of sample preparation which are prone to human error, the accuracy of the fluorescent data captured depends directly on the camera of smartphone used. The accuracy of allergen detection relies on capturing clear and detailed images of the fluorescent signals. The resolution of the smartphone camera used for image capture can impact the sensor's sensitivity and precision. The images captured in such cases largely depends on camera parameters such as ISO and white balance adjustments. While ambient light scattering causes decrease in S/N, low sample concentration is likely to be affected by undesirable noise and most importantly phone's model is a major factor affecting the quality of image captured. Shielding the smartphone camera and the optomechanical enclosure housing different optical components from outside noise (unwanted light signals) can significantly improve the accuracy of the detection system. Additionally, developing customized applications targeting auto-focus and exposure adjustment can help minimizing the effect of using different variants of smartphone models.

Alternatively, these issues can be mitigated by using ALS (Ambient Light Sensor) of the smartphone which is relativity similar for both high end and low-end smartphone model. Thus, in addition to the camera sensor, ALS can also be considered for fluorescence-based estimation of food allergens on a smartphone platform. Thus, with incorporation of custom designed applications employing AI & IoT along with 3D printed optomechanical hardware attachments, smartphone-based fluorescence sensing systems can open new avenues towards low-cost, field-portable and robust hand-held devices addressing the growing demand of food allergen management across the globe.

References

1. G.M. Ross, M.G. Bremer, M.W. Nielen, Consumer-friendly food allergen detection: moving towards smartphone-based immunoassays. Anal. Bioanal. Chem. **410**, 5353–5371 (2018)
2. L. Monaci, E. De Angelis, N. Montemurro, R. Pilolli, Comprehensive overview and recent advances in proteomics MS based methods for food allergens analysis. TrAC, Trends Anal. Chem. **106**, 21–36 (2018)

3. D. Dhruve, V.P. Ratrey, A. Lowanshi, S. Chandravanshi, N. Kashyap, S. Pandey, Comprehensive review of seafood allergens: regulations, detection methods, and implications for public health. J. Aquat. Food Prod. Technol. 1–14 (2024)
4. K. Sheng, H. Jiang, Y. Fang, L. Wang, D. Jiang, Emerging electrochemical biosensing approaches for detection of allergen in food samples: a review. Trends Food Sci. Technol. **121**, 93–104 (2022)
5. X. Weng, S. Neethirajan, A microfluidic biosensor using graphene oxide and aptamer-functionalized quantum dots for peanut allergen detection. Biosens. Bioelectron. **85**, 649–656 (2016)
6. J. Zhou, R. Ai, J. Weng, L. Li, C. Zhou, A. Ma, Y. Wang, A "on-off-on" fluorescence aptasensor using carbon quantum dots and graphene oxide for ultrasensitive detection of the major shellfish allergen Arginine kinase. Microchem. J. **158**, 105171 (2020)
7. J. Zhou, Q. Qi, C. Wang, Y. Qian, G. Liu, Y. Wang, L. Fu, Surface plasmon resonance (SPR) biosensors for food allergen detection in food matrices. Biosens. Bioelectron. **142**, 111449 (2019)
8. M. Khedri, M. Ramezani, H. Rafatpanah, K. Abnous, Detection of food-born allergens with aptamer-based biosensors. TrAC, Trends Anal. Chem. **103**, 126–136 (2018)
9. L. Hong, M. Pan, X. Xie, K. Liu, J. Yang, S. Wang, S. Wang, Aptamer-based fluorescent biosensor for the rapid and sensitive detection of allergens in food matrices. Foods **10**(11), 2598 (2021)
10. D. Calabria, M. Zangheri, S.R.S. Pour, I. Trozzi, A. Pace, E. Lazzarini, M. Guardigli, Luminescent aptamer-based bioassays for sensitive detection of food allergens. Biosensors **12**(8), 644 (2022)
11. W. Su, D. Liang, M. Tan, Microfluidic strategies for sample separation and rapid detection of food allergens. Trends Food Sci. Technol. **110**, 213–225 (2021)
12. H.Y. Lin, C.H. Huang, J. Park, D. Pathania, C.M. Castro, A. Fasano, H. Lee, Integrated magneto-chemical sensor for on-site food allergen detection. ACS Nano **11**(10), 10062–10069 (2017)
13. A.S. Franca, L.S. Oliveira, Applications of smartphones in food analysis, in *Smartphone-Based Detection Devices* (Elsevier, 2021), pp. 249–268
14. J.L. Nelis, S. Moddejongen, X. Guan, A. Anderson, M.L. Colgrave, J.A. Broadbent, Secure food-allergen determination by combining smartphone-based raw image analyses and liquid chromatography–mass spectrometry for the quantification of proteins contained in lateral flow assays. Anal. Chem. **94**(49), 17046–17054 (2022)
15. Y. Qin, S. Zhang, J. Qian, F. Meng, J. Yao, M. Zhang, Lable-free aptamer portable colorimetric smartphone for gliadin detection in food. Front. Bioeng. Biotechnol. **12**, 1338408 (2024)
16. A.F.S. Silva, F.R. Rocha, A novel approach to detect milk adulteration based on the determination of protein content by smartphone-based digital image colorimetry. Food Control **115**, 107299 (2020)
17. C. Xiao, G. Ross, M.W. Nielen, J. Eriksson, G.I. Salentijn, W.C. Mak, A portable smartphone-based imaging surface plasmon resonance biosensor for allergen detection in plant-based milks. Talanta **257**, 124366 (2023)
18. J.V. Kumar, J.W. Rhim, Fluorescent carbon quantum dots for food contaminants detection applications. J. Environ. Chem. Eng. 111999 (2024)
19. Y. Wang, Z. Li, H. Lin, P.N. Siddanakoppalu, J. Zhou, G. Chen, Z. Yu, Quantum-dot-based lateral flow immunoassay for the rapid detection of crustacean major allergen tropomyosin. Food Control **106**, 106714 (2019)
20. T.K. Das, S. Ganguly, Revolutionizing food safety with quantum dot-polymer nanocomposites: from monitoring to sensing applications. Foods **12**(11), 2195 (2023)
21. L.N. Gómez-Arribas, E. Benito-Peña, M.D.C. Hurtado-Sanchez, M.C. Moreno-Bondi, Biosensing based on nanoparticles for food allergens detection. Sensors **18**(4), 1087 (2018)
22. S.M. Saad, J. Abdullah, Biosensors based on II–VI semiconductor quantum dots for health protection, in *Handbook of II-VI Semiconductor-Based Sensors and Radiation Detectors: Vol. 3: Sensors, Biosensors and Radiation Detectors* (Springer International Publishing, Cham, 2023), pp. 633–652

23. Y. Shen, Y. Wei, C. Zhu, J. Cao, D.M. Han, Ratiometric fluorescent signals-driven smartphone-based portable sensors for onsite visual detection of food contaminants. Coord. Chem. Rev. **458**, 214442 (2022)
24. E.M. Khalaf, H.S. Jabbar, R.M. Romero-Parra, G.R.L. Al-Awsi, H.S. Budi, A.S. Altamimi, K.A. Alkhuzai, Smartphone-assisted microfluidic sensor as an intelligent device for on-site determination of food contaminants: developments and applications. Microchem. J. **190**, 108692 (2023)
25. X. Linghu, R. Wang, Y. Lu, Smartphone-integrated fluorescent quenching immunochromatographic test strips designed for the determination of sesame allergens in food samples. Spectrochim. Acta Part A Mol. Biomol. Spectrosc. **280**, 121522 (2022)
26. X. Linghu, J. Qiu, S. Wang, Y. Lu, Fluorescence immunoassay based on magnetic separation and ZnCdSe/ZnS quantum dots as a signal marker for intelligent detection of sesame allergen in foods. Talanta **256**, 124323 (2023)
27. J. Yang, Y. Zhang, L. Wu, Y. Lu, A coffee-ring effect-based paper sensor chip for the determination of beta-lactoglobulin in foods via a smartphone. Sens. Actuators, B Chem. **374**, 132807 (2023)
28. I. Ortiz-Gómez, A. Ipatov, E. Barreiro-Docío, A. Salinas-Castillo, I. de Orbe-Payá, L.F. Capitán-Vallvey, M. Prado, Microfluidic paper-based analytical aptasensor for fluorometric β-lactoglobulin determination. Microchem. J. **198**, 110121 (2024)

Detection of Adulterants in Soy Based Products Using Non-destructive Spectral and Spectral Imaging Techniques

Subarna Deb and Nishant Rachayya Swami Hulle

Abstract Food product adulteration is increasing rapidly, which has an adverse effect on consumer goods' shelf life and safety. Finding food adulterants and toxicants can help the world's population avoid a wide range of harmful health concerns. Soyabean based ingredients are used in large number in the food as well as animal feed industry as a high protein source and have been observed to be adulterated as well as itself used as adulterant in various food products. This chapter discusses the need and techniques used in adulterants detection with regard to Soybean and it's products. The current book chapter focuses on non-thermal spectroscopy and analytical techniques like UV–Vis, Infrared (IR), Raman, Near-Infrared (NIR), Hyperspectral Imaging and MIR spectroscopy. The available literature in the detection of adulterants using spectral and spectral imaging techniques has been discussed.

Keywords UV–Vis · IR · MIR · NIR · FTIR · Hyperspectral imaging · Raman spectroscopy · Food adulteration

1 Introduction

The topic of food safety and it's possible contaminants remains a major factor in gaining consumer acceptance for processed food products [1] After processing operations, the food becomes dense source of energy and nutrients, however, the consumers have the right to believe products to be safe and of highest standard [2]. Food safety is the state in which food is safe to eat, satisfies nutritional needs, and poses no threat to human health. Food adulteration, foreign material doping, food toxicity, pesticide residue, illicit additives, etc. are some of the challenges related to food safety [3]. The safety of food is threatened by food adulteration severely. The problems associated with food adulteration include substitution, diluting, counterfeiting, unapproved additions, production on the grey market or theft diversion, mislabelling, hiding, and so on [4, 5]. Strict monitoring during manufacturing and arbitrary inspection of the

S. Deb · N. R. S. Hulle (✉)
Department of Food Engineering and Technology, Tezpur University, Tezpur, India
e-mail: nishant@tezu.ernet.in

R. Biswas (ed.), *Optical Techniques for Assessing Food Adulterants*, Progress in Optical Science and Photonics 35, https://doi.org/10.1007/978-981-96-8142-6_8

food products in retail sale ensures compliance with regulations. These days, food fraud is a significant ethical and financial issue that impacts the global food sector [6]. Food adulteration is pervasive in a wide range of items, such as packaged meals, alcoholic drinks, jam, milk, meat, honey, fruit juices, vegetable oils and so on [7]. Because consuming good quality food is essential to absorbing the nutrients required for the body to grow and develop to its full potential, food quality inspections and consumer safety regulations have become global mandates [8]. The various factors that affects the quality and safety of food products has been a matter of concern within the food supply chain and it attracted the attentions of government, various regulatory bodies and the researchers [9].

Any kind of detection strategy's main emphasis relies on the type of adulterant, the degree of contamination, and the validation [10]. Conventional methods of food adulteration detection like capillary electrophoresis, ion chromatography, GC–MS, and HPLC, measure the pertinent chemical components and produce reasonably accurate results [11, 12]. In order for these techniques to function, samples must be destroyed, and some of them also require sample pretreatment and hence, large-scale application is not supported by the intricate operating process of detecting the adulterants [13]. As a result, new, non-destructive and quick methods have been developed recently for the assessment of food quality. Few of these methods include spectral imaging, Raman, UV–VIS, and NMR spectroscopy [14].

Soybean (*Glycine max*) is one of the most significant crops that is grown worldwide. Although it originated in East Asia, it has been grown all over the world since the eighteenth century. Asia, North America and South America now grow the majority of the world's soybean [15]. Soya bean is an important crop with significant importance due to its rich oil content (~18%), quality protein content (~40%) [16]. Soyabean accounts for almost two thirds of the world's protein concentrate that is used to feed cattle and 25% of the edible oil produced worldwide. Fish and poultry majorly benefit from the use of soybean meal [17].

Asian cuisines have been using soybeans, as a mainstay for thousands of years. Vegetarian and vegan diets are particularly popular among those who consume soy and soy-based goods due to high-quality protein content and can be processed to make meat and milk alternatives [18]. Soybean is high in protein and vegetable oil [19]. The only legume that has sufficient levels of both alpha-linolenic acid and the necessary omega-3 fatty acid is soybean [20]. Furthermore, compared to fruits, vegetables, and the majority of animal products, soybeans have a higher concentration of biotin, which is acknowledged as a vital nutritional element [21]. Human health benefits from phytochemicals found in soybeans [22–24]. Soybean comprises of 40% proteins, 15% fibers, 15% carbohydrates, and 20% fats. Additionally, they comprises of micronutrients like phosphorus, potassium, magnesium, zinc, copper, vitamin B and C [25].

Some of the soy products can be used as adulterants in food products of higher value such as adulteration of extra-virgin oil with soyabean oil [26]. Maintaining the integrity of the market and protecting consumer safety mainly depend on the quality of the products. This chapter explores the principles, uses, and efficacy of sophisticated spectral and imaging techniques used to identify adulterants with regard to soybean products. Commonly seen adulterations may include addition of cheaper ingredients, use of inferior quality food products, and contamination with harmful substances such as pesticides, herbicides, and industrial contaminants. These practices can lead to allergic reactions, nutritional deficiencies, and toxicity, causing acute or chronic health issues, including organ damage and cancer.

1.1 Challenges in Detecting Adulterants Related to Soy Products

Too much use of Benzoic acid in Soy sauce may cause allergic reactions in body. In addition, Caramel colour being a brown liquid/powder has been used in dark soy sauce as an additive. Some caramel colour contains 4-methylimidazole (4MEI) which is hazardous to human body. The legally added additives to soy sauce are usually regarded as safe as long as they stay within the regulatory parameters. In order to prevent the incorrect use of additives, authenticity analysis must be taken into account concerning the types and quantities of additives utilized [27]. Another kind of adulterant found in soybean products is the presence of Genetically Modified (GMO) soybean in Crude soybean oil [28]. Another very commonly found adulteration in Soy based beverage also known as Soy milk is the addition of water in order to increase the volume and profitability and decrease the nutritional value. Density determination technique, freezing point depression analysis and cryoscopy is used to detect fraud in this case. These conventional techniques, however, can be really time consuming and require specialised tools like a cryoscopic apparatus in case of freezing point depression analysis and densimeter in case of density analysis. Keeping this in view, it can be said that spectroscopic methods may provide a good substitute where such equipment is not accessible [29].

Traditional techniques for detecting adulterants related to Soybean products such as chemical analysis, sensory evaluation and microbiological assays are very accurate but still possess many disadvantages. For example, they demand a large amount of sample preparation which may include intricate sample extraction and purification process that not only lengthens the evaluation time but also demands highly skilled laboratory personnel. Because these approaches are labour intensive and time consuming, they are not suitable for large no. of sample analysis where quality may be compromised [30].

2 Optical Non-destructive Techniques

Optical non-destructive techniques are the analytical techniques used in laboratories and food industries to determine or study the properties of a components, material or the system without causing damaging he samples [31]. Spectroscopic techniques

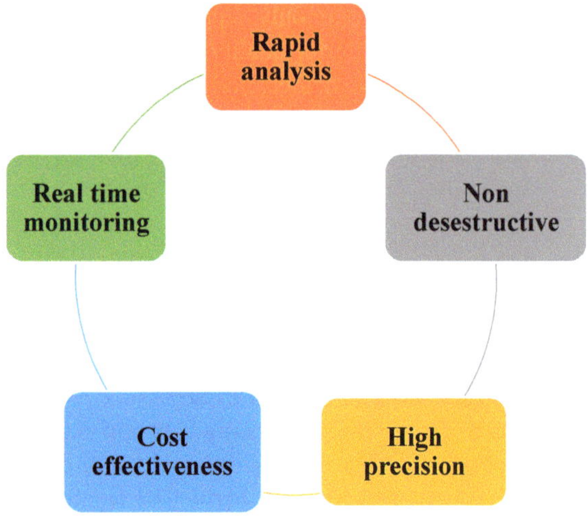

Fig. 1 Advantages of optical techniques

typically takes one or a few minutes and is reasonably accurate and affordable form of analysis with real time monitoring facility (Fig. 1).

2.1 Spectral and Hyperspectral Imaging Techniques

Spectroscopy approaches has proved to be an effective means of analysing and identifying adulterants in various foods products in the recent past. The importance of various Spectral and imaging technique in the process of adulterants detection is growing because of its speed, repeatability, non-destructive qualities, and—most importantly—how simple it is to integrate this cost effective technology into an industrial setting, such as an in-line or on-line application [32]. Spectroscopic techniques are rapidly becoming important analysis tools due to the reproducibility, sensitivity and non-invasive capabilities and in particular due to the easy handling and cost-effective technology in industries [33]. Following are the few spectral techniques commonly used in quality analysis and adulterants detection techniques in relation to Soybean and it's products.

2.1.1 UV–Vis Spectroscopy

According to El Darra et al. [6], UV–Vis spectroscopy is an advanced technique used in laboratories that has high speed, cost effective, and easy to use for food products evaluation. When coupled with chemometrics, the UV–VIS spectra can be a useful tool for food adulteration detection. They consists of both qualitative and quantitative data about the matrix. It can be considered as fast, dependable and cost-effective technique in routine analysis laboratories [34].

Ultraviolet–Visible (UV–Vis) spectroscopy detects the absorbance of UV and visible light by a sample, providing information about its composition. This analytical technique counts the discrete UV wavelengths in nanometres that are absorbed or transmitted by materials based on their concentration and composition. In addition to differences in sample level, the molecular structure of the chemical affects the maxima (λ max) and intensity of the wavelength absorption, whereas other impurities do not have significant effect [35]. It is used to detect melamine in soybean products by measuring absorbance changes at specific wavelengths. It is simple, fast, and suitable for routine analysis. The samples are prepared and exposed to UV–Vis light. The resulting absorbance spectra are analysed to identify and quantify adulterants [36]. The quality of Soybean oil along with few other vegetable oils was analysed for comparison of oxidative stability after heating using UV–Vis spectroscopy coupled with MCR-ALS. This given information about tocopherol degradation in the oil and oxidation products appearance due to the heating process. The results showed that the tocopherol concentration of soybean and oil started to decline and oxidation products started to increase at 50 °C as compared to 70 °C in olive oil. The findings suggested that Soybean and maize oils were less resistant to the increasing temperature as compared to colza, sunflower and olive oils [37].

2.1.2 Infrared (IR) Spectroscopy

Infrared spectroscopy involves the absorption of infrared light by molecular vibrations, producing a characteristic spectrum. It is effective in detecting starch and other carbohydrate adulterants in soybean flour [38]. The samples are analysed using Fourier-transform infrared (FTIR) spectroscopy. The spectra are compared to reference spectra to identify adulterants. In FTIR spectroscopy vibrational spectroscopy uses electromagnetic radiation in order to interact with functional groups present in the samples, that gives different vibrational energy levels [39].

It is quick and non-invasive analytical method which make FT-IR spectroscopy preferrable for the rapid analysis of different products [40]. Nevertheless, the physical and chemical properties of samples may affect the obtained FT-IR spectra and therefore obtained raw data need to be normalized [41]. Four normalizing techniques applicable to FT-IR spectral data include area normalization, minimum–maximum normalization, amide normalization and vector normalization.

Rodionova et al. [42] had performed chemometric non-targeted detection of adulterants in soyabean meal by using infrared spectroscopy. The experiment had presented a rapid and effective systematic approach for detecting food fraud by various adulterants. One-class classification modelling and Near Infrared (NIR) spectroscopic measurements were used in non-targeted analysis as chemometric data processing techniques. It was illustrated by the examination of a numerous set of NIR soybean meal spectra. Both the infected and clean samples were examined. The primary benefit of the suggested method is that it does not try to identify and measure a specific contaminant.

Another experiment was performed in order to determine adulteration by water in Soy based beverages (SSBs) by the use of infrared spectroscopy combined with chemometrics. A collection of 80 SSBs samples were classified using the one class and multiclass approaches applied to obtained spectroscopy data. For one class modelling, technique of data driven SIMCA (DD SIMCA), Soft Independent Modelling Class Analogies (SIMCA), uneven dispersed classes (UNEQ), and one class random forest (OC-RF) were used. Adulterated samples were used as non-targeted class (NT) and non- adulterated samples were used as targeted class (TA) in the models built. For multiclass modelling, techniques such as duel class random forest (DC-RF), partial least square discriminant analysis (PLD-DA), k nearest neighbours (k-NN) were employed. For PLS-DA and k-NN, samples were classified into four groups: (non-adulterated samples, samples adulterated with water at 5% v/v, 10% v/v, and 20% v/v of water) [29].

2.1.3 Raman Spectroscopy

Raman spectroscopy technique is an alternative that offers great accuracy and precision and can offset the drawbacks of NIR spectroscopy, such like its weaker sensitivity for minute components. It offers comprehensive details regarding molecule vibrations. Raman spectroscopy has proven to have exceptionally high sensitivity in a variety of fields [43]. Raman spectroscopy has been found ideal for detection of adulterants in edible oil in many past studies [26, 44]. The benefits of both near-IR and infrared spectroscopy are combined in Raman spectroscopy. Since the spectral vibrational bands determine the chemical and structural bonds within the analyte, Raman spectroscopy has been also utilized to determine the composition of *cis/trans* isomers in edible vegetable oil. One of the most often used quantitative technique is the external standard method (ESM). Using the peak area with associated concentration, a calibration curve is first obtained that can be used to measure the concentration based on the measured sample's peak height or area [45]. The benefits of Raman spectroscopy include: working with both liquids and solids, no need of sample preparation; the non-destructive method and is extremely specific to a material's chemical fingerprint; samples can be examined through glass or polymer packaging; Long-distance optical fibre transmission of Raman scattered light and laser light is possible for remote analysis [46]. Raman spectroscopy is based on the scattering of monochromatic light, providing information about molecular vibrations. It is used for detecting various adulterants, including non-soy proteins and fillers. This technique requires minimal sample preparation and is non-destructive [47]. Baeten et al. [48] used Fourier Transform (FT) Raman spectrometer with multivariate procedures in the detection of adulteration in virgin olive oil adulterated with pomace from olive, soybean oil and corn oils at percentages 1, 5 and 10% respectively. Here, the best result was an adjusted R^2 value of 0.964, analysed by regression on principal components (PCR), that gave 100% accurate discrimination between genuine and adulterated samples. Also, at different adulteration levels; the classification was 91.3% correct.

2.1.4 Near-Infrared (NIR) Spectroscopy

NIR spectroscopy analyses the absorption of near-infrared light by combinations of vibrations and molecular overtones. It is widely used for detecting adulterants in soy flour and soy protein isolates. This process is rapid, non-destructive, and can analyse complex samples [49]. The samples are scanned using NIR spectroscopy, and the resulting spectra are analysed using chemometric techniques to identify adulterants. In the context of detection of adulterants from Soyabean meal, near-infrared (NIR) spectroscopy is frequently utilized because to its rapidity, non-destructiveness, and lack of pollution. It also requires little to no sample preparation [50]. In soybean, NIR Spectroscopy has been used to detect the content of crude protein, oil content, dietary fibre, fatty acids and few physical traits (weight, length, width, volume, density, etc.). It has been also used to predict the leaf soluble protein, and melamine adulteration in soybean meal [51].

In NIR spectroscopy, vibrational energy changes of molecular bonds in chemical compounds result in the observation of NIR absorption signals [52]. Transitions are only permitted between adjacent energy levels in the midinfrared region, while energy transitions across many energy levels take place in the near-infrared region, and so one of the primary types is overtone vibration. Apart from the overtones, a blend of various vibrational transitions might be seen. Due to the likelihood of overtone and the absorption of combination vibrational transitions reduces dramatically with their order. Low intensity bands are associated with overtone and combination vibration [53].

Haughty et al. [33] had experimented on detecting Melamine adulteration in Soybean meal. Here, dehulled (GM and non-GM), hulled, and toasted soya were first contaminated with Melamine followed by creating spectra using NIR Spectroscopy. Chemometric analysis of the spectrum data produced good calibration models and prediction statistics. Additionally, depending on the sample type, data pre-processing method, and mathematical methodology used, the coefficients of determination (R2) were found to range from 0.89 to 0.99. The validation data that was subsequently produced, root mean square error of calibration (0.081–0.276%) and root mean square error of prediction (0.134–0.368%). This suggests that NIRS technique can be utilized to identify fraud and adulteration of soy-based products used as ingredients in animal feed.

2.1.5 Hyperspectral Imaging

Hyperspectral imaging (HSI) has drawn a lot of attention from a variety of fields as a promising non-destructive and quick measurement technique [54]. HSI can give machine vision as well as spectral information at the same time by combining spectroscopy and machine vision technologies into a single device [55] without requiring sample preparation or creating safety risks, hyperspectral imaging (HSI) provides a quick, non-invasive, and affordable technology for non-destructive seed assessment [56]. It allows for a detailed analysis of the moisture content, chemical composition and internal quality. HSI can span a broad range of wavelengths, from 400 to 15,000 nm, on the particular camera being utilized [57–59]. Hyperspectral imaging can be used to identify the characteristics of the sample under examination, as well as their quantity and location [60].

HSI has been employed in a number of experiments to predict seed quality and assess the chemical makeup of seeds, including their protein, fat, and carbohydrate content using spectral information. Using HSI, each chemical component's distinct spectral signature may be found [61]. This technique detects multiple adulterants in soy-based products, such as soy milk and tofu it is useful for comprehensive analysis with spatial and spectral information. The samples are imaged, and the hyperspectral data is processed to identify adulteration patterns based on spectral signatures [47].

For food geographical origin traceability, metabolic profiles, multi-elements, stable isotopes and integrated spectroscopic fingerprints are the priorities. Until now it was difficult to detect adulteration with the same one from several geographic origins than it is to trace it back to a specific geographic origin. In a study conducted by [27] variable selection by uninformative variable elimination using near infrared hyperspectral imaging was followed by the construction of a classification model to distinguish between indigenous and imported soybeans using partial least square discriminant analysis. Thus, this model was able to distinguish between soybeans that were imported and those that were domestic. Additionally, the created model identified 13.3, 20.0, 26.7, and 33.3% of imported soybean as tainted domestic soybean.

2.1.6 MIR Spectroscopy

MIR spectroscopy is a non-destructive technique that has been proved to be useful in the detection of food adulterants through determination of molecular structure, quantifying the intramolecular force and the character of the chemical bonds present in it. It utilizes a beam of light passing through the sample and quantifies its absorbance and transmission [62]. The beneficial features of using MIR spectroscopy for determining food adulterants includes quick analysis as compared to methods such as HPLC; i.e., less than a minute per sample, high speed, easier sample preparation, small sample size, better sensitivity, cost effectiveness etc. It is also capable of analysing sample in different physical state (solid, liquid, gas). However, there are drawbacks of using this technique. One of them being the necessity to use samples that must have IR-active chemical bonds. So, it becomes difficult to detect ions and elements in the sample. Other drawbacks include upgradation of regression model, difficulty in the analysis of aqueous solutions and complex mixtures etc. [63] (Fig. 2).

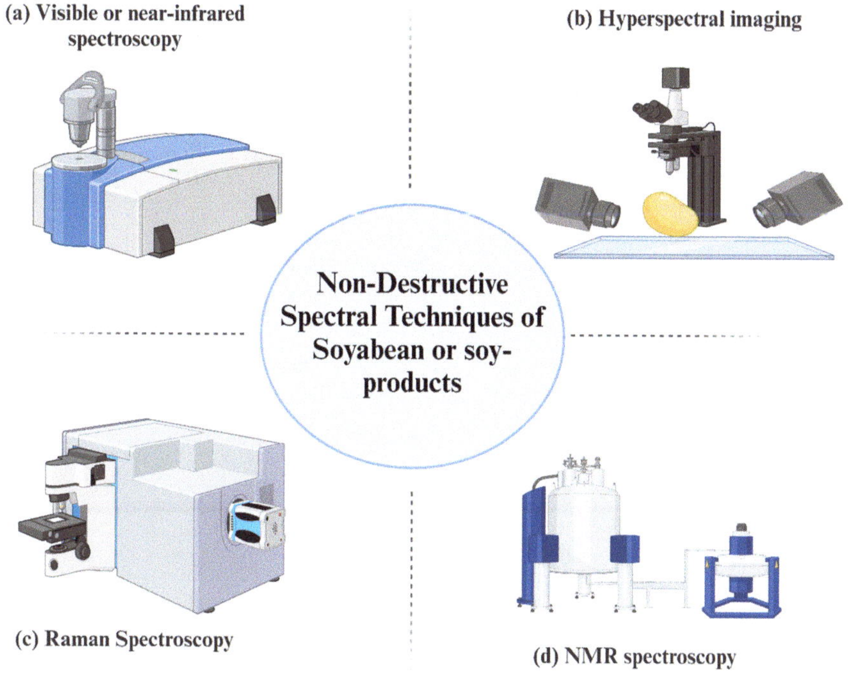

Fig. 2 Non-destructive spectral techniques of determining adulteration

Xue et al. [64] has non-destructively determined the adulteration of Soyabean and Wheat flour in Quinoa flour by using front-face synchronous fluorescence spectroscopy. Investigations were conducted into situations involving single and double adulterants (6–50 wt%). Principal component analysis (PCA) combined with linear discriminant analysis (PCA–LDA) was utilized for initial qualitative discrimination in cases of single adulteration using either soybean or maize flour. Tian et al. [65] had conducted a study in order to identify adulterants in Raw bovine milk i.e., Soymilk using electronic nose and headspace-gas chromatography ion-mobility spectrometry. The study had used headspace-gas chromatography ion-mobility spectrometry (HS-GC-IMS) and electronic nose (E-nose) to identify adulteration in Raw Bovine Milk with Soy Milk rapidly and accurately. Principal component analysis could be used to identify the contaminated samples with Soy Milk using the data acquired from HS-GC-IMS and E-nose (Table 1).

Table 1 Other recent application of emerging technologies with regard to non-destructive evaluation of soybean

Sample	Technique	Applications	Spectral range	Model	References
Soy meal	NIRS	Melamine adulteration	12,000–3800 cm^{-1}	PLS, PCR	[33]
Soybean	NIR-HSI	Detection of adulterated samples	886.3–1693.9 nm	PLS-DA	[27]
Soyabean	MIR and NIR	Quality	3040 and 2850 cm^{-1} 5200 and 5100 cm^{-1}	PLS	[66]
Soybean	NIR	Contaminants	400 and 2498 nm	SIMCA	[42]
Quinoa flour with Soybean flour	FFSFS	Soybean adulteration	$\lambda_{ex} = 250$–500 nm $\Delta\lambda = 30$–180 nm	PCA-LDA	[64]

3 Conclusion

The detection of adulterants in food products is an important aspect of ensuring food quality and safety. Whether intentional or unintentional, adulteration causes severe health risks to the consumers. For instance, melamine or non-soy fillers adulteration in soybean products may lead to allergic reactions and kidney damage highlighting the importance of effective detection methods in order to prevent such risks. Apart from health risks, the economic impact of adulteration in food is profound. The chapter talked about various spectroscopic techniques that are convenient, non- destructive, cost effective, dependable and pollution free. These techniques have proven effective in detecting contaminants with regard to Soy based products. Detecting adulterants in an effective way facilitates traceability in food supply chain enabling quick identification of contaminated products. However, studies should also focus on combining spectral analysis with advanced data processing technologies like machine learning, principle component analysis (PCA), artificial neural networks (ANN), mulivariae statistical analysis etc. In the detection of adulterants in food for more accuracy.

It is also important to notice that these laboratory based optical techniques although play great role in adulterant detection.

In soybean based products, the practical use of such techniques become limited due to the complexity of their size and instrumentation. Consequently, miniaturised and portable sensors such as colorimetric sensors and p-e-nose for analysing food authentication can provide dual benefits of non- destructive as well as real-time, on-site detection of food adulteration. Additionally, combination with AI algorithms can significantly enhance he data analysis capabilities making the process more practical and ensure food quality and integrity.

References

1. Y. He, X. Bai, Q. Xiao, F. Liu, L. Zhou, C. Zhang, Detection of adulteration in food based on nondestructive analysis techniques: a review. Crit. Rev. Food Sci. Nutr. **61**(14), 2351–2371 (2021)
2. F. Fung, H.S. Wang, S. Menon, Food safety in the 21st century. Biomed. J. **41**(2), 88–95 (2018)
3. Y.Z. Feng, D.W. Sun, Application of hyperspectral imaging in food safety inspection and control: a review. Crit. Rev. Food Sci. Nutr. **52**(11), 1039–1058 (2012)
4. G.P. Danezis, A.S. Tsagkaris, V. Brusic, C.A. Georgiou, Food authentication: state of the art and prospects. Curr. Opin. Food Sci. **10**, 22–31 (2016)
5. S. Sumar, H. Ismail, Adulteration of foods–past and present. Nutr. Food Sci. **95**(4), 11–15 (1995)
6. N. El Darra, H.N. Rajha, F. Saleh, R. Al-Oweini, R.G. Maroun, N. Louka, Food fraud detection in commercial pomegranate molasses syrups by UV–VIS spectroscopy, ATR-FTIR spectroscopy and HPLC methods. Food Control **78**, 132–137 (2017)
7. K. Goyal, P. Kumar, K. Verma, Food adulteration detection using artificial intelligence: a systematic review. Arch. Comput. Methods Eng. **29**(1), 397–426 (2022)
8. K. Goyal, P. Kumar, K. Verma, Food adulteration detection using artificial intelligence: a systematic (2021)
9. M. Banti, Food adulteration and some methods of detection, review. Int. J. Nutr. Food Sci. **9**(3), 86–94 (2020)
10. V.S. Raghavan, B. O'Driscoll, J.M. Bloor, B. Li, P. Katare, J. Sethi, D. Jenkins, Emerging graphene-based sensors for the detection of food adulterants and toxicants—a review. Food Chem. **355**, 129547 (2021)
11. B. Chen, B.B. Xu, Z.T. Wang, Determination of eight food additives in food by high performance liquid chromatographic. Food Sci. Technol. **35**(5), 301–306 (2010)
12. Q. Liu, Z. Gong, D. Li, T. Wen, J. Guan, W. Zheng, Rapid and low-cost quantification of adulteration content in camellia oil utilizing UV-vis-NIR spectroscopy combined with feature selection methods. Molecules **28**(16), 5943 (2023)
13. Y. He, X. Bai, Q. Xiao, F. Liu, L. Zhou, C. Zhang, Detection of adulteration in food based on nondestructive analysis techniques: a review. Crit. Rev. Food Sci. Nutr. **61**(14), 2351–2371 (2021). https://doi.org/10.1080/10408398.2020.1777526
14. W.H. Su, H.J. He, D.W. Sun, Non-destructive and rapid evaluation of staple foods quality by using spectroscopic techniques: a review. Crit. Rev. Food Sci. Nutr. **57**(5), 1039–1051 (2017)
15. W. Kudełka, M. Kowalska, M. Popis, Quality of soybean products in terms of essential amino acids composition. Molecules **26**(16), 5071 (2021)
16. R. Modgil, B. Tanwar, A. Goyal, V. Kumar, *Soybean (Glycine max) Oilseeds: Health Attributes and Food Applications* (2021), pp. 1–46
17. D.K. Agarwal, S.D. Billore, A.N. Sharma, B.U. Dupare, S.K. Srivastava, Soybean: introduction, improvement, and utilization in India—problems and prospects. Agric. Res. **2**(4), 293–300 (2013)
18. G. Rizzo, L. Baroni, Soy, soy foods and their role in vegetarian diets. Nutrients **10**(1), 43 (2018)
19. A. Sugiyama, Y. Ueda, H. Takase, K. Yazaki, Do soybeans select specific species of Bradyrhizobium during growth? Commun. Integr. Biol. **8**(1), e992734 (2015)
20. M. Messina, V. Messina, The role of soy in vegetarian diets. Nutrients **2**(8), 855–888 (2010)
21. F. Depeint, W.R. Bruce, N. Shangari, R. Mehta, P.J. O'Brien, Mitochondrial function and toxicity: role of the B vitamin family on mitochondrial energy metabolism. Chem. Biol. Interact. **163**(1–2), 94–112 (2006)
22. A. Crozier, I.B. Jaganath, M.N. Clifford, Dietary phenolics: chemistry, bioavailability and effects on health. Nat. Prod. Rep. **26**(8), 1001–1043 (2009)
23. E.H. Kim, H.M. Ro, S.L. Kim, H.S. Kim, I.M. Chung, Analysis of isoflavone, phenolic, soyasapogenol, and tocopherol compounds in soybean [*Glycine max* (L.) Merrill] germplasms of different seed weights and origins. J. Agric. Food Chem. **60**(23), 6045–6055 (2012)

24. T.S. Tyug, K.N. Prasad, A. Ismail, Antioxidant capacity, phenolics and isoflavones in soybean by-products. Food Chem. **123**(3), 583–589 (2010)
25. C.A. Nogueira-de-Almeida, I.S. Ferraz, F.D.V. Ued, A.C.F. Almeida, L.A.D. Ciampo, Impact of soy consumption on human health: integrative review. Braz. J. Food Technol. **23**, e2019129 (2020)
26. T.O. Mendes, R.A. da Rocha, B.L. Porto, M.A. de Oliveira, V.D.C. dos Anjos, M.J. Bell, Quantification of extra-virgin olive oil adulteration with soybean oil: a comparative study of NIR, MIR, and Raman spectroscopy associated with chemometric approaches. Food Anal. Methods **8**, 2339–2346 (2015)
27. X. Li, D. Wang, L. Yu, F. Ma, X. Wang, D. Pérez-Marín, P. Li, L. Zhang, Origin traceability and adulteration detection of soybean using near infrared hyperspectral imaging. Food Front. **5**(2), 237–244 (2024)
28. Z. Nikolić, I. Vasiljević, G. Zdjelar, V. Đorđević, M. Ignjatov, D. Jovičić, D. Milošević, Detection of genetically modified soybean in crude soybean oil. Food Chem. **145**, 1072–1075 (2014). https://doi.org/10.1016/j.foodchem.2013.09.017
29. E.H. de Paulo, A.M. Rech, F.H. Weiler, M.H. Nascimento, P.R. Filgueiras, M.F. Ferrão, Evaluation of adulteration in soy-based beverages by water addition using chemometrics applied to ATR-FTIR spectroscopy. Food Control **166**, 110746 (2024)
30. R. Goyal, P. Singha, S.K. Singh, Spectroscopic food adulteration detection using machine learning: current challenges and future prospects. Trends Food Sci. Technol. 104377 (2024)
31. P. Kot, M. Muradov, M. Gkantou, G.S. Kamaris, K. Hashim, D. Yeboah, Recent advancements in non-destructive testing techniques for structural health monitoring. Appl. Sci. **11**(6), 2750 (2021)
32. K.M. Sørensen, B. Khakimov, S.B. Engelsen, The use of rapid spectroscopic screening methods to detect adulteration of food raw materials and ingredients. Curr. Opin. Food Sci. **10**, 45–51 (2016)
33. S.A. Haughey, S.F. Graham, E. Cancouët, C.T. Elliott, The application of near-infrared reflectance spectroscopy (NIRS) to detect melamine adulteration of soya bean meal. Food Chem. **136**(3–4), 1557–1561 (2013)
34. R.R. de Souza, D.D. de Sousa Fernandes, P.H.G.D. Diniz, Honey authentication in terms of its adulteration with sugar syrups using UV–Vis spectroscopy and one-class classifiers. Food Chem. **365**, 130467 (2021)
35. M.A. Farag, M. Sheashea, C. Zhao, A.A. Maamoun, UV fingerprinting approaches for quality control analyses of food and functional food coupled to chemometrics: a comprehensive analysis of novel trends and applications. Foods **11**(18), 2867 (2022)
36. K. Kucharska-Ambrożej, J. Karpinska, The application of spectroscopic techniques in combination with chemometrics for detection adulteration of some herbs and spices. Microchem. J. **153**, 104278 (2020)
37. R.P. Gonçalves, P.H. Março, P. Valderrama, Thermal edible oil evaluation by UV–Vis spectroscopy and chemometrics. Food Chem. **163**, 83–86 (2014)
38. R. Pandiselvam, N.U. Sruthi, A. Kumar, A. Kothakota, R. Thirumdas, S.V. Ramesh, D. Cozzolino, Recent applications of vibrational spectroscopic techniques in the grain industry. Food Rev. Intl. **39**(1), 209–239 (2023)
39. R.K. Sen, K. Karthikeyan, P. Prabhakar, J. Vishwakarma, G. Gupta, S.N. Mishra, N. Dwivedi, Fast tracking of adulterants and bacterial contamination in food via Raman and infrared spectroscopies: paving the way for a healthy and safe world. Sens. Diagn. **1**(4), 673–685 (2022)
40. B.J. Lee, Y. Zhou, J.S. Lee, B.K. Shin, J.A. Seo, D. Lee, Y. Kim, H.K. Choi, Discrimination and prediction of the origin of Chinese and Korean soybeans using Fourier transform infrared spectrometry (FT-IR) with multivariate statistical analysis. PLoS ONE **13**(4), e0196315 (2018)
41. X. Zhao, H. Zhu, J. Chen, Q. Ao, FTIR, XRD and SEM analysis of ginger powders with different size. J. Food Process. Preserv. **39**(6), 2017–2026 (2015)
42. O.Y. Rodionova, J.F. Pierna, V. Baeten, A.L. Pomerantsev, Chemometric non-targeted analysis for detection of soybean meal adulteration by near infrared spectroscopy. Food Control **119**, 107459 (2021)

43. H. Lee, B.K. Cho, M.S. Kim, W.H. Lee, J. Tewari, H. Bae, S. Sohn, H.Y. Chi, Prediction of crude protein and oil content of soybeans using Raman spectroscopy. Sens. Actuators, B Chem. **185**, 694–700 (2013)
44. W. Dong, Y. Zhang, B. Zhang, X. Wang, Quantitative analysis of adulteration of extra virgin olive oil using Raman spectroscopy improved by Bayesian framework least squares support vector machines. Anal. Methods **4**(9), 2772–2777 (2012)
45. X.F. Zhang, M.Q. Zou, X.H. Qi, F. Liu, C. Zhang, F. Yin, Quantitative detection of adulterated olive oil by Raman spectroscopy and chemometrics. J. Raman Spectrosc. **42**(9), 1784–1788 (2011)
46. K.M. Khan, H. Krishna, S.K. Majumder, P.K. Gupta, Detection of urea adulteration in milk using near-infrared Raman spectroscopy. Food Anal. Anal. Methods **8**, 93–102 (2015)
47. W. Li, F. Tan, J. Cui, B. Ma, Fast identification of soybean varieties using Raman spectroscopy. Vib. Spectrosc. **123**, 103447 (2022)
48. V. Baeten, M. Meurens, M.T. Morales, R. Aparicio, Detection of virgin olive oil adulteration by Fourier transform Raman spectroscopy. J. Agric. Food Chem. **44**(8), 2225–2230 (1996)
49. H. Cen, Y. He, Theory and application of near infrared reflectance spectroscopy in determination of food quality. Trends Food Sci. Technol. **18**(2), 72–83 (2007)
50. G. Shen, J.A.F. Pierna, V. Baeten, Y. Cao, L. Han, Z. Yang, Local anomaly detection and quantitative analysis of contaminants in soybean meal using near infrared imaging: the example of non-protein nitrogen. Spectrochim. Acta Part A Mol. Biomol. Spectrosc. **225**, 117494 (2020)
51. R. Xu, W. Hu, Y. Zhou, X. Zhang, S. Xu, Q. Guo, J. Wang, Use of near-infrared spectroscopy for the rapid evaluation of soybean [*Glycine max* (L.) Merri.] water soluble protein content. Spectrochimica Acta Part A: Mol. Biomol. Spectrosc. **224**, 117400 (2020)
52. Y. Saito, T. Suzuki, N. Kondo, Evaluation of optical properties of tofu samples produced with different coagulation temperatures and times using near-infrared transmission spectroscopy. Infrared Phys. Technol. **123**, 104149 (2022)
53. L. Chen, Z. Yang, L. Han, A review on the use of near-infrared spectroscopy for analyzing feed protein materials. Appl. Spectrosc. Rev. **48**(7), 509–522 (2013)
54. L. Wang, L. Pang, L. Yan, J. Zhang, Nondestructive rapid identification of soybean varieties using hyperspectral imaging technology. J. Appl. Spectrosc. **89**(1), 84–91 (2022)
55. Y. Zhao, C. Zhang, S. Zhu, P. Gao, L. Feng, Y. He, Non-destructive and rapid variety discrimination and visualization of single grape seed using near-infrared hyperspectral imaging technique and multivariate analysis. Molecules **23**(6), 1352 (2018)
56. S.W. Jeong, J.I. Lyu, H. Jeong, J. Baek, J.K. Moon, C. Lee, M. Choi, Y.I. Park, SUnSeT: spectral unmixing of hyperspectral images for phenotyping soybean seed traits. Plant Cell Rep. **43**(7), 1–17 (2024)
57. X. Gao, T. Wang, J. Yang, J. Tao, Y. Qiu, Y. Meng, B. Mao, Y. Li, Deep-learning-based hyperspectral imaging through a RGB camera. J. Electron. Imaging **30**(5), 053014–053014 (2021)
58. C.T. Kucha, L. Liu, M. Ngadi, G. Claude, Hyperspectral imaging and chemometrics as a non-invasive tool to discriminate and analyze iodine value of pork fat. Food Control **127**, 108145 (2021)
59. L.D. Medus, M. Saban, J.V. Francés-Víllora, M. Bataller-Mompeán, A. Rosado-Muñoz, Hyperspectral image classification using CNN: application to industrial food packaging. Food Control **125**, 107962 (2021)
60. M. Kamruzzaman, D.W. Sun, G. ElMasry, P. Allen, Fast detection and visualization of minced lamb meat adulteration using NIR hyperspectral imaging and multivariate image analysis. Talanta **103**, 130–136 (2013)
61. C. Erkinbaev, K. Henderson, J. Paliwal, Discrimination of gluten-free oats from contaminants using near infrared hyperspectral imaging technique. Food Control **80**, 197–203 (2017)
62. W.H. Su, D.W. Sun, Mid-infrared (MIR) spectroscopy for quality analysis of liquid foods. Food Eng. Rev. **11**, 142–158 (2019)
63. S. Bureau, D. Cozzolino, C.J. Clark, Contributions of Fourier-transform mid infrared (FT-MIR) spectroscopy to the study of fruit and vegetables: a review. Postharvest Biol. Technol. **148**, 1–14 (2019)

64. S.S. Xue, J. Tan, J.Y. Xie, M.F. Li, Rapid, simultaneous and non-destructive determination of maize flour and soybean flour adulterated in quinoa flour by front-face synchronous fluorescence spectroscopy. Food Control **130**, 108329 (2021)
65. H. Tian, J. Xiong, S. Chen, H. Yu, C. Chen, J. Huang, H. Yuan, X. Lou, Rapid identification of adulteration in raw bovine milk with soymilk by electronic nose and headspace-gas chromatography ion-mobility spectrometry. Food Chem. X **18**, 100696 (2023)
66. D.S. Ferreira, O.F. Galão, J.A.L. Pallone, R.J. Poppi, Comparison and application of near-infrared (NIR) and mid-infrared (MIR) spectroscopy for determination of quality parameters in soybean samples. Food Control **35**(1), 227–232 (2014)

A Study on the Detection of Toxicity and Adulteration in Food Products and Proposing a Simple Detection System Using Metal Oxide Semiconductor (MOS) Based Electronic Nose

Nimisha Dutta, Shama Tanveer, and Kaushik Dehingia

Abstract Food safety is undeniably one of the major worldwide challenges, and in the present times gas sensors and e-nose systems are used to overcome this challenge as they are far less expensive, more efficient and easier to miniaturise as compared to other available options. As a result, real-time monitoring can be improved in food manufacturing and processing industry for the detection of toxicity and adulteration, by the use of such systems. However, in most cases we have observed that the devices or the technologies that are provided for the detection of toxicity and adulteration in food products, are not only expensive, but are way too far from the reach of the common household. Hence, under such situations it becomes immensely important to work with an aim to develop more efficient and cost-effective systems or device that can be used for the detection of toxicity and adulteration in food products, which at the same time is accessible to all parts of the society equally. The potential of Metal Oxide Semiconductor (MOS)-based e-nose systems as economical and effective substitutes for real-time food safety monitoring is examined in this chapter. The chapter is organised into four key sections. The basic ideas and workings of MOS gas sensors and e-nose systems are presented in the first section, emphasising how well suited they are for pollutant detection. The widespread problem of food adulteration is covered in the second section, which also describes how e-nose systems may detect different types of adulterants. Using MOS technology, the third portion suggests a revolutionary toxicity detection framework that is simple, inexpensive, and widely applicable. Future developments to improve e-nose systems' sensitivity, selectivity, and usefulness in identifying a variety of pollutants in food and pharmaceutical goods are covered in the last section. By bridging the gap between advanced technologies and practical accessibility, this chapter lays the groundwork for developing innovative, low-cost solutions for food safety. It underscores the importance of democratising access to reliable detection systems, ensuring equitable food safety for all segments of society.

N. Dutta (✉) · S. Tanveer · K. Dehingia
Department of Physics, Dibrugarh University, Dibrugarh, Assam, India
e-mail: nimishadutta@dibru.ac.in

R. Biswas (ed.), *Optical Techniques for Assessing Food Adulterants*, Progress in Optical Science and Photonics 35, https://doi.org/10.1007/978-981-96-8142-6_9

Keywords MOS sensor · E-nose · Food adulteration · ANN

1 Introduction

1.1 Gas Sensors

Gas sensors are such an electronic devices design to detect and quantify the presence of various gases, including oxygen, carbon dioxide, nitrogen, methane, etc. They are integral to systems used for detection of harmful gas leaks and monitoring air quality industrial, commercial, and official settings. Widely utilised in the manufacturing and factory environments, gas sensors ensure safety by identifying leaks and measuring the concentration of the gas. Their incorporation into security and embedded systems further aids in detecting hazardous gas levels. They are calibrated more often than other types of sensors and emit audible and visual signals when dangerous levels of gases are detected. Originally developed for use in coal mines to accurately monitor hazardous gases, their applications have diversified to include the chemical industry, environmental pollution monitoring, and healthcare sectors. In an optimised gas sensor design, the power consumption and heat losses must be minimised [1]. A Typical human nose has 400 different types of scent receptors enabling us to smell about 1 trillion different odours. But still, many of us do not have the capacity to identify the type or concentration of gas present in our atmosphere. In this aspect, the sensors play a very important role. There are many types of sensors to measure different parameters and a "Gas sensor" is the one which comes handy in applications where we have to detect the variation in the concentration of toxic gases in order to maintain the system safety and avoid or prevent any unexpected threats to human life and environment. There are various gas sensors to detect gases like oxygen, carbon dioxide, nitrogen, methane etc. They are also frequently employed in equipment that monitors the air quality in workplaces and businesses and detects the leakage of dangerous gases. These sensors also measure the amount of gas present in the atmosphere or local environment. They are frequently employed to gauge gas concentrations and identify explosive or dangerous gases. Gas sensors are used in manufacturing facilities and factories to find gas leaks and to detect smoke and carbon monoxide in residential buildings. The size (portable and fixed), range, and sensing capabilities of the gas sensors vary greatly. Gas sensors require more frequent calibration than many other types of sensors since they are continually reacting with air and other gases [1, 2].

1.2 Types of Gas Sensors

The gas sensors are broadly classified into 4 types:

1. Electro-acoustic sensors
2. Field effect transistor (FET)
3. Electrochemical gas sensor
4. Chemiresistive sensors

Among electro-acoustic sensors, the quartz crystal microbalance (QCM) is widely employed in electronic nose (E-nose) systems. QCM-based gas sensors operate using Sauerbrey's equation, which correlates the adsorption and desorption of gas molecules on the transducer's mass to shifts in resonant frequency. QCM based E-nose systems offer several advantages such as real time measurements and fast response, however, there is still room for improvement in terms of sensitivity, specificity, and portability [3].

FET-based sensors responds to target gases by inducing charge transfer or altering the conductive channel, thereby altering the electrical characteristics (mobility and voltage). Electrochemical gas sensors, commonly integrated into external circuits to measure the concentration of a target gas. These sensors are advantageous for their resistance to interference from other gases and their robustness against condensation. However, they are limited by a relatively short lifespan and the need for frequent maintenance [3]. Chemiresistive gas sensors, operating according to Ohm's Law, rely on a catalytic reaction between the sensing material and the target gas. This interaction alters the resistance within the sensor, generating a corresponding voltage signal. Among the various methodologies, Chemiresistive is frequently preferred due to benefits such as ease of fabrication, high sensitivity, and low cost, and is the focus of this review [3].

1.2.1 Chemiresistive Gas Sensors

Chemiresistive sensors typically used two-electrode system in which the resistance of the sensing layer in between the electrodes is measured as a function of time after gas exposure. This sensing layer can be made of a variety of materials, including conducting polymers and metal oxide semiconductor sensors [4].

Conducting Polymer Sensors

Conducting polymer sensors utilise active materials from a conducting polymer family, such as polypyrroles, thiophenes, indoles, or furans. Under the influence of various gases, these polymer sensors change in size and conductance to varying degrees. The polymer material is electro-polymerised between two electrodes with a gap of 10–20 μm. Odourants alter molecular bonding, influencing electron transfer

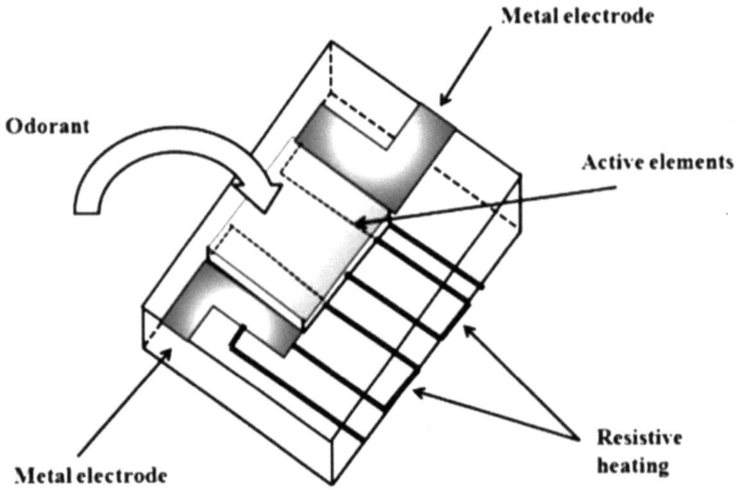

Fig. 1 Conducting polymer sensor

along the polymer chain, thereby changing the sensor's conductivity. Conducting polymer sensors can operate at room temperature and do not require heaters, as MOS gas sensors. However, a significant drawback of conducting polymer sensors is the complexity and time-intensive nature of their manufacturing process, which is highly sensitive to humidity. Figure 1 illustrates a conducting polymer sensor.

Metal Oxide Semiconductor (MOS) Sensors

Metal Oxide Semiconductor (MOS) gas sensors have garnered significant attention due to advancements in MOS nanostructures, including nanoparticles, nanotubes, and nanorods. The sensing mechanism of MOS gas sensors is based on the variation of electrical conductance of the sensing layers in response to target gas exposure. MOS sensors are typically constructed using single or combined metal oxides, such as SnO_2, TiO_2, In_2O_3, WO_3, NiO, and ZnO [5]. A standard MOS gas sensor consists of four key components: (i) Sensing layer, (ii) Insulating substrate, (iii) two contact electrodes, and (iv) heating element. An electrically insulating layer separates the heater from the sensing layer and electrodes. The sensor's sensitivity arises from mechanisms such as electron trapping at adsorbed molecules and band bending induced by these molecules, which alter conductance. Molecular oxygen adsorbed onto the grain boundaries of the MOS layer traps free electrons, forming a depletion layer and bending the energy band[1]. At elevated temperatures (typically >300 °C), oxygen absorbed on the metal oxide reacts with the target gas, altering the resistance of the MOS layer. N-type metal oxides, such as tin oxide and zinc oxide, are commonly used due to their ability to absorb gases. Since these metal oxides can be produced using semiconductor processes, MOS gas sensors are cost-effective and suitable for mass production. However, MOS sensors are less effective for detecting chemically inert and stable gases, such as CO_2, limiting the range of detectable gases[5, 6]. They also consume a significant amount pf power due to their high operating temperatures, however their reactions slow down at lower temperatures. Despite these limitations, MOS gas sensors, also known as "Chemiresistors," are widely used in industry due to their low cost, scalability, and ability to detect a broad range of gases. The Taguchi Gas Sensor (TGS) developed by N. Taguchi and manufactured by Figaro Engineering, TGS sensors operate by allowing electrical current to pass through the grain boundaries of SnO_2 micro-crystals. At these boundaries, oxygen adsorption forms a potential barrier that alters the conductance [5, 7, 8]. MOS-based gas sensors operate at high temperatures ranging from 300 to 500 degrees Celsius, which increases their sensitivity. Despite the fact that MOS sensors are less selective, they have been used in the design of electronic nose systems due to their small size, low cost, and ease of availability. The most appealing feature of these gas sensors is their sensitivity to a wide range of gases and their usable life span of 3–5 years depending on the sensor's usage. A few drawbacks of the sensors however, include a dependency between external humidity and temperature effects to the rate of reaction to a gas and high power consumption. These drawbacks can be overcome by using appropriate signal pre-processing techniques before employing the pattern recognition algorithm [9].

1.3 Electronic Nose (E-nose)

The electronic nose is one of the new innovative method for food quality control which mimics the human sense of smell and offers a fast and cost-effective assessments of food freshness, process monitoring, and quality control. It uses sensor arrays to detect odours in the headspace of samples offering a non-invasive approach to quality evaluation. An E-nose is composed of a sample collector, a sensor array, a data processor, and a pattern recognition system. The sample collector gathers VOCs from the sample and delivers them to the sensor array. The sensor array converts these chemical signals into corresponding electrical signals. To enhance the selectivity of MOS-based gas sensors, data processors and advanced pattern recognition techniques are employed. The electronic nose is inspired by the human sense of smell and has been used as an effective tool in the quality assessment of food products. This system consists of a sensor array, a signal processing subsystem, and a pattern recognition module. The sensor array of the electronic nose detects odours, including large amounts of various volatile compounds in the headspace of the sample, after which the output is provided; representing the fingerprint of all sample components. The "fingerprint" described by electronic nose sensors is used to extract potential information about samples based on a suitable algorithm. The electronic nose is an automated non-destructive method, which has become a popular system due to different advantages such as convenient construction, high sensitivity, and cost effectiveness [6, 7, 10].

Electronic noses are particularly useful for analysing and detecting VOCs due to the cross-reactivity of e-nose multisensory arrays and sensitivity to a diversity of volatile compounds from a wide range of organic chemical classes. These gas-sensing devices have a long history of effective use with many applications previously developed in numerous industries, including agriculture and forestry, food quality and safety, biomedical and forensic sciences, pharmaceutical and drug development, human health and disease diagnoses, and environmental protection [7].

1.3.1 Basic Elements of an E-nose Electronic Noses Include Three Major Parts

(1) A sample delivery system
(2) A detection system.
(3) A computing system.

Sampling system: It creates headspace and injects a sample into the detection system. Many of the devices use Artificial Intelligence to evaluate the detection system signal and run it through an Artificial Neural Network (ANN), which simulates how the human brain works.

Detection system (Sensor Array): It reacts to the presence of VOCs, undergoes chemical change, and generates a usable signal. The adsorption of volatile compounds on the sensor surface changes the sensor's physical properties.

Computing system (Pattern Classifier): It analyses detection system data and produces pattern recognition output that describes the odour or aroma. Combines individual sensor element responses and generated sample data [7, 11, 12].

1.3.2 Application of MOS Based E-Noses in Food Industry

Gas sensors based on MOS are commercially available. They have been more widely used to make e-nose arrays for odour measurement than any other class of gas sensors. Table 1 lists some of the commercially available gas sensors of SnO_2 and ZnO that are manufactured by Figaro Engineering Inc. (Japan), HANWEI ELETRONICS CO., LTD [7, 8, 12].

1.3.3 Biological Olfaction

A gas detecting instrument with a single sensor element features the ability to provide measurement results immediately on-site and to conduct continuous monitoring. However, such sensors are limited in their functionality; they they can measure the strength or ppm level only of a single odour but lack the ability to classify odours

Table 1 Sensor type and their properties

Sensor type	Detection range	Usage area	Advantages	Disadvantages
Metal oxide	5–500 ppm	Food and beverage industry Indoor and outdoor monitoring	Suitable to range of gases Operation in high temperature High power consumption Fast response, small size, easy to use	High sensitivity and specificity Sulphur poisoning Weak precision Humidity sensitive
Conducting polymer	0.1–100 ppm	Medical industry Pharmaceutical industry Food and beverage industry Environmental monitoring	Sensitive to range of gases Fast response and low cost Resistant to sensor poisoning Use at room temperature	Humidity sensitive Temperature sensitive Limited sensor life Affected from drift
Quartz crystal microbalance	1.5 Hz/ppm Ing mass change	Pharmaceutical industry Environmental monitoring Food industry Security systems	Good sensitivity Low detection limits Fast response	Hard to implement Poor signal-to-noise ratio Humidity sensitive Temperature sensitive
Acoustic wave	100–400 MHz	Environmental monitoring Food and beverage industry Chemical detection Automotive industry	Small size and low cost Good sensitivity and response time Response to nearly all gases	Hard to implement Temperature sensitive Poor signal-to-noise ratio
Electrochemical	0–1000 ppm adjustable	Security systems Industrial applications Medical applications	Power efficient and robust High range operation temperature Sensitive to diverse gases	Large size Limited sensitivity

(continued)

Table 1 (continued)

Sensor type	Detection range	Usage area	Advantages	Disadvantages
Catalytic bead	Large scale	Environmental monitoring Chemical monitoring	Fast response High specificity for combustible gases	Operate in high temperature Only for compounds with oxygen
Optical	Change with light parameters Low ppb	Biomedical applications Environmental monitoring	Low cost and light weight Immune to electromagnetic interference Rapid and very high sensitivity	Hard to implement Low portability Affected by light interference

within a group. To achieve sensitivity comparable to the human nose, an E-Nose sensor must possess intelligent classification and discrimination capabilities.

Over the last few decades, our understanding of biological olfaction has grown rapidly, and it led its efforts to replicate this process in artificial systems. Even though the human nose is traditionally used to evaluate odours such as tea, wine, or perfumes, it is subjective and biased by nature. Furthermore, traditional human testing is impractical for industrial applications where dangerous components are present in the gas or when it is necessary to monitor certain volatile compounds for extended periods of time. In such cases, E-Nose is an alternative technique [7, 13, 14].

A chemical sensor array replaces the olfactory receptor cells in the artificial olfactory system. Using chemical sensors in an array with pattern recognition capability provides a higher degree of selectivity, opening up a wide range of applications. The basic idea behind E-Nose is that each odour leaves a unique pattern or fingerprint on each sensor array. The number and type of sensors in the sensor array have a large impact on the degree of selectivity and the type of odours that can be detected. As shown in Fig. 2 the headspace generation ensures the concentration of volatiles of organic solvents before and during sampling. The sensors are exposed to a constant flow of gases through pipelines inside the electronic nose during the collection period, and the response begins. During the purging operation, sensor heads are cleared with a fresh air blast, causing the sensors to return to their baseline values [4].

The E-nose system, comprises of three functional components that works in tandem-

(i) An integrated chemical sensor array,
(ii) An interfacing electronic circuitry,
(iii) A pattern recognition (PARC) software paradigm.

Sensor electronics convert the chemical signal to an electrical signal, amplifying and conditioning it. To digitise and store the response signal for processing, interfacing electronics circuitry is used. Patterns from known odours are used to create a

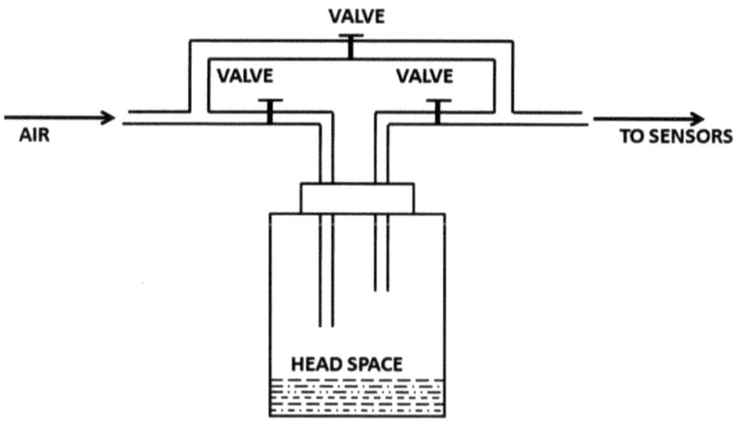

Fig. 2 Showing the air flow inside the sensor heads

database and train a pattern recognition system so that unknown odours can be classified and identified in the future. An E-Nose, as the human olfactory system, discriminates new patterns and associates them to specific odours by training the system using a set of known data samples. Sample delivery systems are used to transfer the odour from the source (typically by a miniature pump) to a sensor chamber in which an array of selected gas sensors is installed [4].

1.3.4 The Process of Odour Delivery Can be Summarised as Follows

The odour delivery system drives each sensor to a known reference state at the start of a sampling process by applying a reference gas (for example, fresh air) to the sensor chamber. The sensor responses in this state are referred to as baseline level. The delivery system then exposes the sensors to a specific odour, causing a transient response as the compounds begin to interact with the surface and bulk of the sensor's active material. After a few seconds to a few minutes, the sensors reach a steady state, at which point the odour is removed from the system and the sensor is returned to its baseline level by pumping the reference gas to prepare the system for a new measurement cycle.2 The above mentioned steps are known in the E-Nose literature as a "three phase sampling process" and they are usually carried out in chambers where humidity, temperature and exposure to the analyte are controlled. Figure 3 shows a typical response of a given array of gas sensors to a three-phase sampling process. Fig. 4 shows the comparison between the three basic elements that comprise an electronic nose and a human nose. Another important consideration is that the sensors' power consumption be kept to a minimum. Other electrical equipment, such as pumps and valves, which frequently share the same power supply, are likely to be present in the headspace containing the sensor array. Furthermore, the smaller the

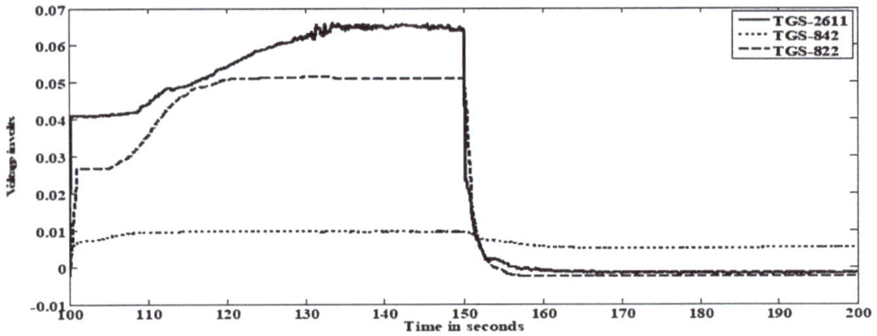

Fig. 3 Typical response of a given array of gas sensors to a three-phase sampling process

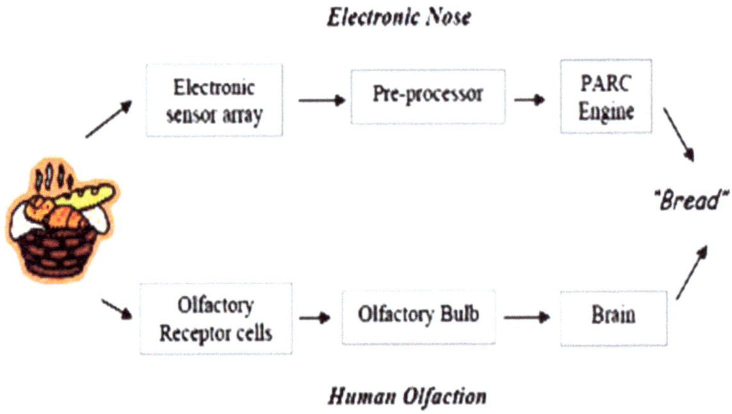

Fig. 4 Comparison between the three basic elements that comprise e-nose and human olfaction

sensor size, the greater the possibility of sensor integration in a variety of platforms, including portable E-Nose [4].

1.4 Signal Processing and Pattern Recognition

MOS gas sensors perform odour classification and discrimination using algorithms supported by learning, training, and decision making. These algorithms are essentially data and signal processing algorithms that operate sequentially. The four stages of E-Nose signal processing and pattern recognition are as follows:

(a) Pre-processing
(b) Feature extraction
(c) Classification
(d) Decision making

An electronic circuit processes the sensor response, translating physical changes such as variations in resistance for chemo-resistive and polymer sensors or changes in mass and resonance frequency for piezoelectric sensors into corresponding voltage or current signals. One critical factor to consider during this stage is sensor drift. Normalising the sensor output is essential, as the voltage levels of output signals can vary due to differences in the sensitivities of the sensors within the array.

The process of classifying involves separating specific samples from a larger sample space. Based on the principles of flexible algorithms, classification aids in sorting samples into their closest matching class or even assigns them to their exact class. The classification stage is used to determine the patterns that are typical of each odour once the sensor response data are projected on an appropriate low-dimensional space. By comparing the identified patterns with those stored during training, the classification step assigns a class label to the data, effectively identifying the odourant [4]. The decision-making stage makes judgment to the classification and even determines that the unknown sample "doesn't belong" or "it belongs" or "it nearly belongs to" any one of the databases.

Electronic nose, or e-nose, technology is used to identify and detect complex odours of VOC emissions from biological systems. In order to detect volatile gases or vapours released by vegetative tissues of a biological sample, this analytical equipment is composed of a number of sensors placed in an array. The e-nose works as a multisensory detector to sense VOC emissions by measuring the changes in electrical conductivity or resistance of individual sensors caused by a signal response to adsorption of different chemical classes of VOCs to the surface sensor coatings due to interactions between the semiconductor and analyte gas molecules. Varying types of volatile sample types cause different responses from the array's sensors to the sample scents (VOC emissions). In order to distinguish between variations in sensor data responses to VOCs detected in the plant sample headspace, the output responses from all the sensors are then sent to computer system records. Using multivariate data analysis methods, this enables the classification and identification of unidentified aromas [4, 7, 8, 15]. Data analysis by an E-nose can be represented in four stages: data acquisition, feature extraction, classification, and decision making. Figure 5 shows the block diagram representing signal processing architecture in an E-nose.

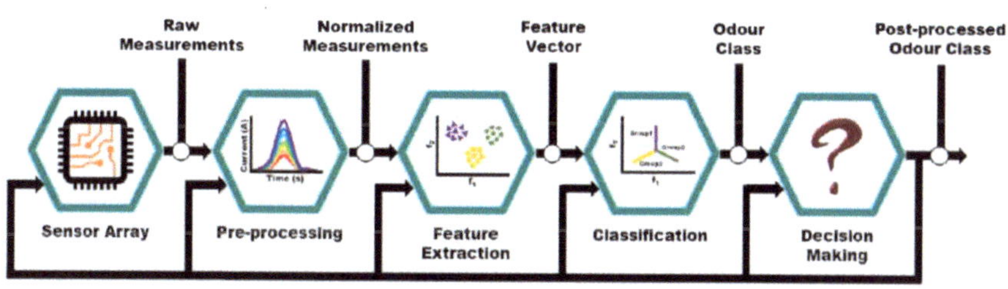

Fig. 5 Block diagram representing signal processing architecture in an E-nose

1.4.1 Data Acquisition

The process of digitising data from the environment so that it may be presented, examined, and saved in a computer is known as data acquisition, or DAQ as it is frequently abbreviated. It is the process of sampling signals that measure real world physical conditions and converting the resulting samples into digital numeric values that can be manipulated by a computer.

Components of a Data Acquisition System

The three components Sensor, Signal Conditioning, and Analog-to-Digital Converter make up every data acquisition system (ADC).

1. Sensors: It converts physical parameters to electrical signals.
2. Signal conditioning circuit: It converts sensor output into a form that can be converted to digital values.
3. Analog-to-digital converters (ADC): It converts conditioned sensor signals to digital values [16].

1.4.2 Pattern Recognition

Pattern recognition is the process of finding unique patterns using a machine learning algorithm. The information collected from these discovered patterns can subsequently be used for classification. In other words, classification is an example of pattern recognition in which a trained model divides the data into classes. The pattern recognition algorithms of ENs utilise a variety of learning-based techniques, which are crucial for enhancing EN systems. Some frequently used tools include principle component analysis (PCA), support vector machines (SVMs), artificial neural networks, stepwise discriminant analysis (SDA), partial least squares regression (PLSR), generalised least squares regression (GLSR), multiple linear regression (MLR), and discriminant function analysis (DFA), Artificial Neural Networks (ANNs). The most popular data processing for various EN sensor types are listed in Table 2, and algorithm characteristics, including benefits and drawbacks, are provided below [17].

Principle Component Analysis (PCA)

PCA is a well-known and often used unsupervised method for linearly reducing the dataset's dimensionality. The primary components of the original variables' highest variance are used to represent the high-dimensional data in a new, lower-dimensional subspace. The eigenvectors of the data covariance matrix are these primary components. The most significant contribution among dimensions is made

Table 2 E-nose sensor type and data analysis methods

Sensor type	Data analysis method (s)
Metal oxide	PCA, ANNs, discrimination analysis, regression
Conducting polymer	PCA, ANNs, discrimination analysis
Quartz crystal microbalance	PCA, ANNs, discrimination analysis
Acoustics wave	PCA, ANNs, discrimination analysis
Electro-chemical	Pattern recognition methods
Catalytic bead	PCA
Optical	PCA

by the highest eigenvalue in magnitude and its related eigenvector, which preserve the most variance. A lot of EN applications employ PCA [17].

Linear Discriminant Analysis (LDA)

LDA is a method of linear transformation used to reduce the number of dimensions. In contrast to PCA, LDA is a supervised approach that aims to maximise class discrimination while taking class labels into account which performs better for bigger multiple-class [7].

Support Vector Machines (SVMs)

SVMs can be used to solve classification and regression problems. They are frequently used in linear and nonlinear binary classification problems as a supervised learning technique. SVMs essentially choose the best-fitting hyper plane, or decision boundary, to help classify data points clearly [13].

Artificial Neural Networks (ANNs) and Machine Learning (ML) Algorithms

Machine learning (ML) was first used to describe the ability of a computer to learn without being specifically programmed for a specific purpose in 1959 [7]. There are four types of learning: supervised learning, semi-supervised learning, unsupervised learning, and reinforcement learning [17]. In supervised learning, the ML algorithm is given a dataset with known (labelled) inputs and outputs (labels). A reliable mapping function (or model) between inputs and outputs is what is desired. ML methods that are semi-supervised can be used to speed up processing and minimise computing complexity in problems involving a lot of unlabeled input. Unsupervised learning makes use of a dataset that only includes inputs. In order to extract underlying structures from the input data, the ML algorithm probes the dataset, seeking out specific

Fig. 6 General block diagram of machine learning algorithms

patterns by examining correlations and interactions between data points. Unsupervised learning is used to solve problems in dimensionality reduction, association, and clustering. The processes of data gathering, modelling, training, and evaluation make up a general block diagram of machine learning shown in Fig. 6.

In any machine learning (ML) study, obtaining a relevant dataset is essential. Without a dataset associated with the specific context, achieving the desired successful solution becomes impossible. Thus, a pre-existing or specially constructed dataset is always a prerequisite. Once a suitable dataset that meets the particulars of the study and its information, ML methods are used to create a model. Selecting the right parameters during the modeling stage is crucial to achieving high accuracy. Following the training process, the system can be assessed using a variety of techniques, including cross-validation [10].

The operational concepts of biological neural networks serve as inspiration for ANNs. These networks are typically designed as fully connected, multilayer structures. Where the number of hidden layers used depends on the specific task. The activations of the hidden layers are influenced by the input layer and the weights connecting them. Similarly, the output layer's activation is determined by the weights between the hidden layers. It is important to note that ANNs' gates are the activation functions of individual neurons since without them, every network would act like a linear regression model. Despite the existence of linear functions, nonlinear activations are typically preferred to help the network acquire intricate data structures and intricate functional mappings in order to produce accurate predictions. Another reason to choose nonlinear functions is that they are more suitable to be used in back-propagation [10].

Therefore, when trained with datasets comprising sufficient and reasonable samples, ANNs and their latest developments have been a popular alternative in a variety of EN applications. Trained artificial neural networks (ANNs) can quickly recognise distinct odours in a chemical mixture as data is supplied into the system. But throughout the training process, a robust system must be built using a lot of datasets. It is crucial to use a paradigm in a gas sensor array that can combine separate sensor outputs and offer a thorough identification and classification result. Additionally, there have been a lot of research recently on EN technology combined with ANNs in a variety of application sectors [10].

1.4.3 Features Extraction

The process of turning raw data into numerical features that can be processed while keeping the information in the original data set is known as feature extraction.

Compared to using machine learning on the raw data directly, it produces better results.

2 Adulteration and Detection of Toxic Elements in Food Products Using MOS Gas Sensors

Adulteration in food has been a concern since the beginning of civilisation, as it not only decreases the quality of food products but also results in a number of ill effects on health. Authentic testing of food and adulterant detection of various food products is required for value assessment and to assure consumer protection against fraudulent activities Food authenticity evaluation is becoming a great source of concern, as it deceives consumers and can constitute a major health risk in some cases. Unfortunately, food safety/integrity issues occur on an alarmingly regular basis, necessitating the development of novel analytical procedures. Food is considered to be adulterated when a substance that debases the quality of food is added to it to make it visually pleasing or preserve it and in turn, makes it hazardous or of lesser nutritional value. Adulteration of food also takes place when cheaper or low-quality substances are used as a substitute for whole and good quality ingredients which in turn reduces the quality of food. Food adulteration has become a widespread issue in the food processing industry. Under various statutes and laws in most countries across the world, the food standards and safety regulatory authority has mandated compositional criteria for practically all food products in order to assure the validity of the product. Unfortunately, in order to make more money, companies often adulterate food with cheap replacements. Adulteration with poisonous chemical like formalin and calcium carbide is widespread and regularly applied on fish [18], meat [19], fruit [20], dairy products [21] that causes different types of human diseases. Other harmful adulterants that impact various human body parts include colouring colours, urea, and other approved preservatives [2, 11].

2.1 *Harmful Effects of Food Adulteration on Human Health*

Food Adulteration has a great negative impact on our health. Irrespective of the type of adulteration prolonged consumption of adulterated food is very harmful to the human body. Consuming adulterated food increases the number of toxic elements in the body. Nutritional value also falls as more and more adulterants are added to the food thus leaving it the least nutritive for the body. Chemical adulterants and colours which are added to food many times prove to be fatal as they pose an onset of health risks or hazards because of the high percentage of carcinogens present in them. Some adulterated food may also affect our organs directly and without much delay by leading to heart, kidney, liver, and many more organ disorders and failure. Food adulteration risk-including food defense incidents, intended to harm or commit food fraud motivated by economic gain is growing in scope, scale, and threat. The adulteration risks are direct, indirect, and technical. Examples of fraud include melamine or chalk in milk, carcinogen colours or dyes in foods, dilution of fruit juices, species swapping of seafood, unauthorised repackaging, tax-avoidance smuggling, or the sale of stolen goods [22].

It is critical to monitor food products throughout the food supply chain in order to protect processed foods and ensure an appropriate level of quality. Some methods for detecting food adulteration include microscopic examination, differential scanning calorimetry, gas chromatography-mass spectrometry, Fourier transform infrared spectroscopy, high performance liquid chromatography, nuclear magnetic resonance, sensory, and chemical detection. Most approaches for frequent use in the food sector are expensive, adulterating compound specialised, and time consuming, necessitating a high level of technical understanding in these disciplines. As a result, real-time rapid detection systems for quality control, particularly rapid adulteration detection in processed foods are being commercialised. Techniques based on e-nose, which approximates the human nose, are commonly employed to deal with adulteration issues in food products for this purpose [1, 11, 23].

2.2 *Types of Food Adulteration*

Mainly, there are 4 different types of food adulteration and those are as follows:

1. Packing Hazard—The packing material in which the food is packed may also contaminate and deteriorate the food's ingredients, resulting in packing hazards.
2. Intentional Adulteration—When similar-looking substances are added to a food item in order to increase its weight or quantity and profit. For example, dilatation of milk with water, adding sand or pebbles to pulses, and so on.
3. Metallic Adulteration—Metallic adulteration is the intentional or unintentional addition of metallic materials such as lead or mercury to food.
4. Incidental Adulteration—This refers to residues of pesticides or harmful fertilisers, the presence of rodent droppings, larvae growth, and other factors that occur as a result of carelessness when handling food.

History of Food Adulteration and methods of food adulteration go hand in hand as with evolving time the methods of food adulteration also evolved and updated. Some of the methods of food adulteration which are still practiced today are:

- Artificial ripening entails adding chemicals to fruits and vegetables to hasten the ripening process, which is also considered food adulteration because it can alter the nutritional value of the food.
- Mixing—The mixing of inedible items such as sand, clay, dust, pebbles, and so on with edible items.

- Misbranding entails changing the manufacturing dates, expiry dates, ingredient list, or misleading consumers by not providing the full list of ingredients' derivatives, among other things.
- Substitution entails replacing healthy and high-quality ingredients with cheaper, lower-quality, and potentially hazardous alternatives that may also pose a health risk [24].
- Addition of Toxic Substances—Food adulteration also includes the addition of substances to food in order to increase profits through increased sales. Example—Addition of colours and dyes which aren't edible, harmful or unpermitted, or unsafe preservatives.
- Using Decomposed Food—This method involves mixing decomposed food with healthy ones. Food that covers up any damage or inferiority of any manner is also known as adulterated food as this intentional adulteration of good and healthy food with decomposed one leads to food adulteration [11].

2.3 *Major Food Adulteration Events*

1. Brick dust is added in chilli powder on the other hand lead chromate and metanil yellow is added in turmeric powder that deteriorates health condition
2. Sugar and salt is adulterated with calcium carbonate, similarly urea is mixed in parched rice.
3. The saw dust is added in coriander powder and cumin powder.
4. Fresh vegetables and fruits, the main part of our meal is adulterated by the self-applied fertilisers and pesticides by the farmers to increase the production and to avoid the attacks of insects and diseases.
5. Vegetables and fruits are also openly sold at roads which cause the free fallen smoke and dust particles resulting in contamination; appear to occur with alarming regularity, and there is no doubt that public concern about this will grow in tandem with population pressures [24].

As a result, there is a growing need for round-the-clock vigilance, which necessitates the research and development of rapid analytical and detection techniques in the field of food authenticity assessment. In this regard, two approaches, namely electronic noses (e-noses) and electronic tongues, are emerging as promising tools in the attempt to efficiently address this issue [25] (e-tongues). Both are sensor systems, but when applied to a given liquid sample, they do not look at the same features; the former is in contact with its headspace, whereas the latter is immersed in the sample [26].

2.4 *Application in Food Adulteration*

Detection of adulteration in various foods include fruits, alcoholic and non-alcoholic drinks, milk and dairy products, honey, meat, fish, spices, edible oils, tea, and coffee. The evolutions of e-noses in quality monitoring of foodstuffs particularly for falsification detection has been summarised below in Table 3 [13].

2.4.1 Ghee

The cow ghee is derived from milk and contains high-quality nutrients. Ghee is considered a memory enhancer, grasping booster and controlling senses and strengthening them [49, 50]. However, ghee is more expensive than edible oils due to the scarcity of ghee for consumption and high demand during the lean season. As a result, ghee producers frequently tamper with it by adding less expensive edible oils, hydrogenated fat, animal body fat, and starches. Food adulteration is the process by which the authenticity of any high-value food molecule is diminished using less expensive materials of the same type. (Traders frequently adulterate cow ghee to increase their profits.) The conventional techniques for detecting adulteration need sample pre-processing, which takes time and money. In order to quickly and accurately assess the quality of goods, machine olfaction equipment are widely used today. To detect adulteration in ghee with hydrogenated fat, a prospective electronic nose (E-nose) system based on metal oxide semiconductor (MOS) gas sensors was developed (vanaspati). Using multivariate chemometric analysis, such as principal component analysis (PCA) and discriminant function analysis (DFA) approach for adulteration detection in ghee, the data produced from the e-nose system was examined for pattern identification and classification. In the e-nose dataset, the PCA described 98.10 percent of the variation whereas the DFA explained 99.10 percent. The accuracy in training data and its cross validation was found to be 98.18 and 97.27%, respectively. The success rate of adulteration identification for test samples using DFA model was 90.90% based on received e-nose signals. The results of PCA and DFA suggest that the developed e-nose system successfully identified pure and adulterated ghee samples based on the e-nose data.

Various methods for detection of adulteration in ghee includes physicochemical methods [51, 52] chromatographic methods [53] and spectroscopic methods [21].

These approaches' shortcomings include high costs, long preparation times, and the need for technical know-how for standard laboratory analysis [5, 19]. The aforementioned limitations highlight the need for more biological olfactory devices to assess food quality. It is possible to detect odourants from any food products using machine olfaction or an electronic nose that closely resembles the human nose in a quicker, more accurate, and more affordable manner with no sample pre-processing and simple handling [13, 50]. E-noses are incredibly helpful for various industrial applications in the sectors of food, pharmaceutics, clinical diagnosis, and environment monitoring [54]. This is because flavour characteristics can be used to assess food quality [49]. The most widely used sensors in e-nose systems are surface acoustic wave (SAW), conducting polymer (CP) and metal oxide semiconductor (MOS), among which MOS sensors are extensively used due to their high chemical stability, low response to moisture, long life, and reasonable price [54, 55] MOS type gas sensors are extensively used in MQ (Hanwei Electronics Co., Ltd., China) and

Table 3 Application of E-nose for monitoring adulteration of food

Category	Product	Adulteration with	Type of e-nose configuration for adulteration monitoring	Detection result to recommend best technique	Reference
Edible oils	Camellia seed oil and sesame oil	Maize oil	Commercial type, Airsense PEN2 (Germany) e-nose system with 10 MOS sensors	LDA: 83.6% for camellia seed oil while and 94.5% for sesame oil	[21]
	Sesame oil	Maize oil	Commercial type, PEN2 (Airsense corporation, Germany) e-nose with 10 MOS sensors	LDA: only one sample is incorrectly classified out of one hundred sixty five samples	[27]
	Virgin olive oil	Sunflower oil and olive-pomace oil	Commercial type, two array chambers of Alpha MOS e-nose (FOX 3000) with 12 (6 + 6) MOS based sensors	LDA: classification accuracy >96% and in some cases almost 100%	[28]
	Olive oil	Hazelnut oil	Commercial type, Alpha MOS e-nose (Fox 4000) system with 3 chambers each containing six MOS sensors	PCA: 96% separation of samples PLS: correlation coefficient 0.997	[29]
	Argan oil	Sunflower oil	Experimental type, e-nose system with 5 MOS gas sensors	PCA: 98.81% in comestible argan oil adulteration and 98.09% in cosmetic argan oil adulteration	[30]
	Soybean oil	Old frying oil	Experimental type, 8 MOS gas sensor based e-nose system	PLS: correlation coefficient of 0.843 for old frying oil adulteration of soybean oil	[21]

(continued)

Table 3 (continued)

Category	Product	Adulteration with	Type of e-nose configuration for adulteration monitoring	Detection result to recommend best technique	Reference
	Virgin coconut oil	Palm kernel olein oil	Commercial type, zNose™ (Model 7100 Electronic Sensor Technology, Newbury Park, CA, USA) e-nose with SAW sensor	PLS: correlation coefficient of 0.91 for palm olein oil adulteration of virgin coconut oil	[31]
	Palm olein oil	Palm stearin oil	Commercial type, MS-E-nose (SMart Nose300; SMart Nose, MarinEpagnier, Switzerland)	DFA: coefficient of determination for DF1 is 0.997 and DF2 is 0.966. DFA could able to discriminate below 10% adulteration level	[32]
	Peony seed oil	Soybean oil, corn oil, sunflower oil, rapeseed oil	Commercial type, PEN3 e-nose (Airsense analytics, Germany) with 10 MOS chemical gas sensors	LDA: soybean oil and sunflower oil can be successfully discriminated from peony seed oil	[33]
Milk and dairy products	Milk	Water and reconstituted milk powder	Commercial type, PEN2 e-nose (WMA Airsense Analysentechnik, Germany) with 10 MOS sensors	LDA: adulterated samples can be i thin discriminated from milk stored for 1–4 days	[34]
	Milk	Aqueous CH, O, H202, NaClO	Experimental type, e-nose device based on 10 MOS gas sensors	SVM: 94.64, 92.85 and 87.75% for aqueous CH2O, H202, NaClO adulteration in milk, respectively	[35]

(continued)

Table 3 (continued)

Category	Product	Adulteration with	Type of e-nose configuration for adulteration monitoring	Detection result to recommend best technique	Reference
	Milk	Detergent powder	Experimental type, e-nose device based on 10 MOS gas sensors	SVM: 94.64, 92.85 and 87.75% for aqueous CH2O, H202, NaCIO adulteration in milk, respectively	[35]
	Ghee	Sunflower oil and cow body fat	Experimental type, e-nose system with 8 different MOS gas sensors	PCA: 96 and 97% accuracy for sunflower oil and cow body fat adulteration of ghee, respectively	[36]
Honey	Honey	Corn syrup and rice syrup	Commercial type, GC-Heracles e-nose (Alpha MOS, Toulouse, France)	PCA and PLS: Both the classifiers couldn't well classified the syrup adulterations in honey	[16]
	Honey	Beetroot sugar and cane sugar	Commercial type, Cyranose 320 (Pasadena, CA, USA) e-nose system with 32 polymer sensors	LDA: 76.5 and 74.9% accuracy for stepwise LDA and direct LDA respectively	[37]
	Honey	Beetroot sugar and cane sugar	Commercial type, Cyranose 320 (Pasadena, CA, USA) e-nose system with 32 polymer sensors	ANN: mean absolute error of 6.9% for fusion data while 15% for e-nose separately	[38]
	Honey	2 different brands of sage syrup	Commercial type, Cyranose 320 e-nose system with 32 polymer matrix sensors blended with carbon black	PNN: 92.59% classification accuracy	[39]

(continued)

Table 3 (continued)

Category	Product	Adulteration with	Type of e-nose configuration for adulteration monitoring	Detection result to recommend best technique	Reference
Alcoholic and nonalcoholic drinks	Liquor	Different geographical spirits	Commercial type, Alpha MOS based SA-Flash GC e-nose (HERACLES II, France)	DFA: 100% classification accuracy	[40]
	Italian wine	Ethanol, methanol, and other brands of wine	Experimental type, e-nose based on 4 MOS for headspace analysis	ANN: 93% of correct classification	[41]
	Fruit juice	Alcohol	Commercial type, Cyranose 320 e-nose system with 32 carbon/polymer matrix sensors	ANN, SVM: apple juice-Raki mixture, 98.33% LDA: lemon juice-alcohol mixture, 95% ANN: orange juice-vodka mixture, 96.67% KNN: sour cherry juice-alcohol mixture, 91.67%	[42]
	Cherry tomato juice	Overripe tomato juices	Commercial type, PEN2 e-nose (Airsense Analytics, Schwerin, Germany) chamber with 10 MOS sensors	PCR: R7 > 0.99 for both training and prediction set of data	[42]
	Fresh cheery tomato juice	Overripe tomato juices upto 30%	Commercial type, PEN2 e-nose (Airsense Analytics, Schwerin, Germany) with 10 MOS positioned in a small chamber	SVM: 100% classification accuracy	[43]

(continued)

Table 3 (continued)

Category	Product	Adulteration with	Type of e-nose configuration for adulteration monitoring	Detection result to recommend best technique	Reference
Meat	Mutton	Duck meat	Commercial type, PEN2 e-nose (airsense Corporation, Germany) with 10 MOS sensor positioned in sensor array	LDA:98.2% classification accuracy	[44]
	Lard	Chicken fat	Commercial type, PEN3 e-nose (Airsense Corporation, Germany) with 10 MOS sensors in sensor array	LDA: 98.2% classification accuracy	[45]
Spices	Spice mixture	Curry and garlic	Commercial type, e-nose (KAMINA-type, Yson GmbH) contains a chip array of 38 sensor segments based on gas sensitive doped tin oxide	PCA: 93.8% classification accuracy	[27]
	Saffron	Aroma fingerprints of saffron with yellow styles, safflower and dyed corn stigma	Experimental type, e-nose system with 6 MOS sensors (HANWEI Electronics Co., Ltd., Henan, China)	ANN: 86.87 and 100% accuracy for saffron adulteration with yellow styles-dyed corn stigma and safflower, respectively	[46]

(continued)

Table 3 (continued)

Category	Product	Adulteration with	Type of e-nose configuration for adulteration monitoring	Detection result to recommend best technique	Reference
	Cumin	Coriander	Experimental type, e-nose system with 6 MOS sensors	PCA: 91.68% classification accuracy	[47]
	Saffron	Artificially colored safflower and artificially colored yellow styles of saffron	Experimental type, e-nose system with 7 MOS sensors	ANN: correlation coefficient of 0.95 and 0.97 for color and aroma based adulteration, respectively SVM: 100% success rate only on aroma dataset	[48]

TGS (Figaro Engineering Inc., Japan) series. A review of recent researches on food adulteration detection using MOS gas sensor based e-nose can be mentioned in the article [13].

2.4.2 Edible Oil

In the human diet, edible plant oils play a significant role in the preparation, cooking, and preservation of food. Some oils are more nutrient-dense and cost more because of their superior quality and flavour. To make a profit, the vendors therefore mix these pricey oils with cheaper oils, fats, and sterols that have a comparable fatty acid profile. In some instances, tainted edible plant oil poses a major health risk [55]. However, an e-nose is a reliable and extremely sensitive method by which a customer can confirm the purity and authenticity of edible oils. Several studies have employed various approaches to identify such adulterations [56, 57].

The virgin olive oil and refined olive oil usually adulterated with crude hazelnut oil was inspected using a direct coupling headspace mass spectrometer chemsensor 4440 system [58, 59] composed of head space auto sampler and a mass spectrometer [11, 13, 60, 61]. Hazelnut oil and olive oil share similar sorts of chemicals, such as total sterols, fatty acid chains, and triglycerols, making it difficult to identify between the two using standard testing methods. Other adulterants that are regularly found in olive oils include rapeseed oil [59], maize oil [61], sunflower oil [9], and olive pomace oil [57]. In order to generate the pattern of the VOCs present in the falsified virgin olive oils with sunflower oils and olive pomace oils, [56] used 12 MOS sensors [12, 13].

2.4.3 Honey

When sugar concentrate is added to honey, a frequent method of adulteration occurs; sadly, scientific testing have not been successful in identifying this fraud. Using a FOX 4000TM E-nose, a Chinese research team[68] has created a technique for the quick identification of adulterated Acacia honey. PCA was used to assess the information (from the e-nose) on both natural and contaminated honey. Samples were spiked with various concentrations of rape honey and rice syrup. The ability of the e-nose to qualitatively detect counterfeit honey was investigated using LDA. According to the findings, the level of adulteration and e-nose signals are correlated linearly. On the other hand, the minimum rape honey and rice syrup additions required to alter the honey scent system were 2 percent and 1 percent, respectively. As a result, adulterant molecules have a simple way of altering the honey scent system. This illustrated the e-potent nose's capacity to distinguish adulterated honey from pure honey. The findings showed that LDA pattern recognition algorithms can distinguish between genuine honey and contaminated honey, which enables quick and precise detection of honey adulteration [58].

3 Evaluation and Modeling of the Device

Many works have been done for the betterment of the society and development of the new sensing technologies, which primly focuses on detection of toxicity present in different types of products that we use every day in our daily lives. Although these methods and technologies (as mentioned in our previous sections) are very much effective and efficient for detection of toxicity in many food products, however, in order to use or apply these methods or technologies for detection of toxicity, the apparatus that we have to use is not only costly but the arrangement of those system and apparatus will be very time consuming. Moreover, using such costly apparatus will deviate us from the main aim of creating a cost effective system for detection of toxicity in daily food products, which can be easily made accessible to even the small household of the nation. For the creation of any toxicity measuring device as we mentioned above, we first need to focus on two main tasks - firstly, the creation of a cost effective vacuum chamber and secondly the creation or selection of the gas sensor array for making the sensing part or the E-nose, for the device. The vacuum chamber will be needed to place the test sample for testing in an ambient, clean and neutral environment. Coming to the E-nose, it will be working as the main working part of our detection system. The E-nose will specifically be the major working part of our device or system, however after that we also have to focus on how effectively we can use this system with more modification and perfections for better results in future. But, at present creation of a cost effective E-nose comes out to be our prime objective.

3.1 Vacuum Chambers

A vacuum is defined as "the space in which there is no matter or where the pressure is so low that any particles in the space have no effect on any processes occurring in the volume." It is a condition that is well below normal atmospheric pressure and is measured in pressure units (pascal). A vacuum can be created by sucking air out of a space with a vacuum pump or by lowering pressure with a fast flowing fluid, as in Bernoulli's principle." Vacuum chambers are pressure vessels that use pumps to extract air and other gases from the chamber, lowering the pressure as it approaches a near absolute vacuum. The majority of commercially available vacuum chambers are constructed from aluminum as seen in the Fig. 7 [58].

The stresses induced on a vacuum chamber are directly proportional to the chamber's surface area. As the volume of a vacuum chamber increases so does its surface area, leading to higher stresses caused by the external atmospheric pressure acting on the chamber walls. This presents a critical challenge in designing a vacuum chamber to withstand atmospheric pressure, which causes stress on the chamber's walls [59].

Vacuum chambers play a pivotal role in enabling research under low-pressure and vacuum conditions. This enables experimentation in conditions that would otherwise be prohibitively expensive and inconvenient, such as sending equipment high into the atmosphere or into low earth orbit. For instance, sending a small 10-pound payload of experimental and data acquisition equipment into orbit can cost upwards of $500,000. Moreover, conducting experiments at such altitudes often limits the ability to modify or adjust the experimental setup in real-time. Vacuum chambers address these challenges, offering a convenient and cost-effective alternative by allowing such experiments to be performed within the confines of a laboratory. To replicate the atmospheric conditions of low Earth orbit, the pressure inside the vacuum chamber must reach levels as low as 10^8 Pascal. The Table 4 provides a comparison of the

Fig. 7 A common vacuum chamber model

Table 4 Vacuum gauge pressure scale

% Vacuum	Torr (mm mercury)	Micron	psia, (Ib/in^2 abs)	Inches mercury absolute	Inches mercury gauge	kPa abs
0.0	760.0	7,60,000	14.7	29.92	0.00	101.4
1.3	750.0	7,50,000	14.5	29.5	0.42	99.9
1.9	735.6	7,35,600	14.2	28.9	1.02	97.7
7.9	700.0	7,00,000	13.5	27.6	2.32	93.5
21.0	600.0	6,00,000	11.6	23.6	6.32	79.9
34.0	500.0	5,00,000	9.7	19.7	10.22	66.7
47.0	400.0	4,00,000	7.7	15.7	14.22	53.2
50.0	380.0	3,80,000	7.3	15.0	14.92	50.8
61.0	300.0	3,00,000	5.8	11.8	18.12	40
74.0	200.0	2,00,000	3.9	7.85	22.07	26.6
87.0	100.0	1,00,000	1.93	3.94	25.98	13.3
88.0	90.0	90,000	1.74	3.54	26.38	12
89.5	80.0	80,000	1.55	3.15	26.77	10.7
90.8	70.0	70,000	1.35	2.76	27.16	9.3
92.1	60.0	60,000	1.16	2.36	27.56	8
93.0	51.7	51,700	1.00	2.03	27.89	6.9
93.5	50.0	50,000	0.97	1.97	27.95	6.7
94.8	40.0	40,000	0.77	1.57	28.35	5.3
96.1	30.0	30,000	0.58	1.18	28.74	4
96.6	25.4	25,400	0.49	1.00	28.92	3.4
97.4	20.0	20,000	0.39	0.785	29.14	2.7
98.7	10.0	10,000	0.193	0.394	29.53	1.3
99.0	7.6	7,600	0.147	0.299	29.62	1.0
99.87	1.0	1,000	0.01934	0.03937	29.88	0.13
99.90	0.75	750	0.0145	0.0295	29.89	0.1
99.99	0.10	100	0.00193	0.00394	29.916	0.013
99.999	0.01	10	0.000193	0.000394	29.9196	0.0013
100	0.00	0	0	0	29.92	0

percentage of absolute vacuum relative to various pressure units. This reference is critical for determining the vacuum gauge pressure achieved within the chamber and assessing the percentage of absolute vacuum obtained during future experimental procedures [60].

Vacuum chambers not only significantly reduce the costs associated of conducting an experiments but also provide a safer environment for testing. At high altitudes or in low Earth orbit, unforeseen irregularities can affect test equipment, compromise results, and disrupt studies in unexpected ways. While validated trials in real-world conditions are required before applying to a final design, preliminary testing in controlled environments is equally important [60–62].

Using a vacuum chamber allows scientists to simulate various conditions, such as those at high altitudes or in low Earth orbit, while saving time, money, and effort. These chambers also offer the flexibility to easily modify sensors and fixtures, enabling researchers to conduct a wide range of experiments efficiently. This adaptability further enhances the utility of vacuum chambers in experimental setups.

From this discussion, it is evident that vacuum chambers play a vital role in our setup, particularly in the context of our toxicity-measuring device. However, it is crucial to note that the construction of the vacuum chamber needs to be done with greatest care and precision for best results. Moreover, as our primary objective is to create a cost-effective device, careful consideration must be given to the costs associated with constructing the vacuum chamber and selecting supporting equipment, such as the vacuum pump. This approach is essential to maintaining the device's affordability and economic viability. For example, one of the authors while working with MQ gas sensors[59], had simply use a glass beaker covered with foil for creating the environment of a clean empty chamber, which we can see in Fig. 8a. Also, again another author while working with ghee adulteration[60], had used a vacuum chamber as shown in the given Fig. 8b respectively. And, from this we can clearly see that the construction of the vacuum.

chamber if desired can be done very cost- effectively, if we know the exact needs and requirements that it needs to satisfy and the limit to which it should satisfy. Hence, coming to the vacuum chamber that we will be constructing in future, mainly should have the following specifications for satisfying the basic needs of our device:

1. Small in size and shape so that it can fit anywhere.
2. Light weight so that it can be easy to carry.
3. Should be able to create a clean and pure vacuum environment to a considerable level.
4. Should consume low power.
5. And, most importantly it should be cost effective in nature and functioning

3.2 E-nose

Now, as discussed in the earlier, the E-Nose, or electronic nose, is essentially an artificial replication of the human nose, designed to mimic its core functions with precision and accuracy when constructed effectively. The e-nose primarily comprises an array of gas sensors, with MOS gas sensors being preferred due to their high selectivity, efficiency, and affordability. The specific type and number of gas sensors to be used in the array when developing the e-nose entirely depends on the type and nature of the target substances that one is dealing with.

(a)

(b)

Fig. 8 **a** Picture of a system used by an author while working with MQ gas sensors. **b** Picture of a system used by an author while working with MQ gas sensors

For instance, in one study [63], the detection of ghee adulteration was successfully carried out using four simple MQ gas sensors, as illustrated in Fig. 9.

Now, coming to our case, for the creation of the respective e-nose of our toxicity measuring device we will be focusing on the following points:

1. Compact Design: The e-nose should be lightweight and occupy minimal space.
2. High Sensitivity and Selectivity: The sensor array must detect a wide range of toxic substances with a minimal number of sensors, requiring the selection of highly sensitive and selective sensors.
3. Energy Efficiency: The e-nose should consume low power to ensure cost-effective and sustainable operation for both the user and the system.

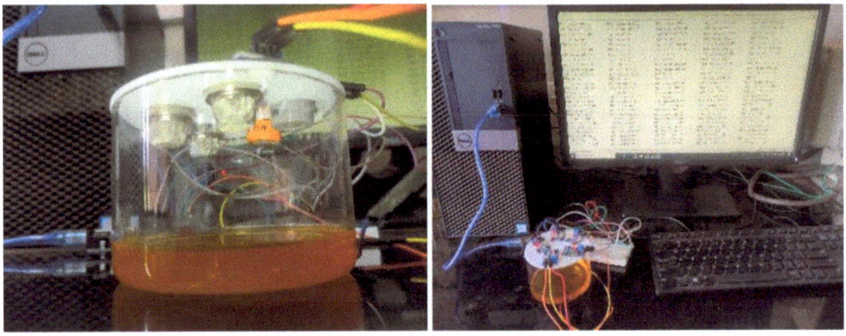

Fig. 9 Picture of a system used by an author while studying adulteration of ghee

4. Market Standards Compliance: It should be capable of detecting a sufficient range of toxic substances in food samples to meet market standards.
5. Cost Effectiveness: The construction of the e-nose should align with the overall goal of creating an affordable device.

However, in our case for creation of the cost effective toxicity measuring device, we will be using the MQ series gas sensors, mainly because of their high efficiency and cost effective nature. And, also they are very easy to use because of the already provided system and environment that we can use with the Arduino-uno micro-controller. Now, coming to the specific type of MQ gas sensors that we are going to use, we will choose them by focusing on the specific type of toxicity that we want to detect.

Regarding the specific MQ sensors for the array, their selection will be guided by the types of toxic substances targeted for detection. Based on the analysis and findings presented in earlier discussion, we have identified a few common toxic substances or gases commonly released by adulterated, damaged, or expired food products. These include:

1. Alcohol
2. VOCs (such as benzene, aldehyde, ketone, and ester)
3. Ammonia

To efficiently detect these toxic gases, we will incorporate MQ sensors that are best suited for their detection, ensuring the e-nose meets its intended purpose of identifying food.

toxicity effectively and affordably.

Form the above Table 5 that is provided by every manufacturer of MQ gas sensors, which shows the respective target gas that is detected by each MQ gas sensor effectively, and by looking at the above table and then matching the names of the respective toxic gases that that we had mention above, we can see that the following MQ gas sensors can be used for the construction of the gas sensor array or e-nose that we need:

Table 5 Different MQ gas sensors and their target gas

Sensor name	Main target gas
MQ-2	Methane, Butane, LPG, Smoke
MQ-3	Alcohol, Ethanol, Smoke
MO-4	Methane, CNG Gas
MQ-5	Natural gas, LPG
MQ-6	LPG, butane
MQ-7	Carbon monoxide
MQ-8	Hydrogen gas
MQ-9	Carbon monoxide, flammable gasses
MQ131	Ozone
MQ135	Air quality
MQ136	Hydrogen sulphide gas
MQ137	Ammonia
MQ138	Benzene, Toluene, Alcohol, Propane, Formaldehyde gas, Hydrogen
MQ214	Methane, Natural gas
MQ216	Natural gas, Coal gas
MQ303A	Alcohol, Ethanol, smoke
MQ306A	LPG, butane
MQ307A	Carbon monoxide
MQ309A	Carbon monoxide, flammable gas

1. MQ-3
2. MQ-3
3. MQ-137
4. MQ-138

Where, the MQ-3 and MQ -303 gas sensors helps in the detection of the alcohol, MQ 137 gas sensor help us in the detection of the ammonia, and also, MQ-138 gas sensor help us in the detection of the VOC respectively. However, we know that for the better detection of toxicity in food products, we not only have to use direct detection system, as mentioned in the above part, but we also have to make use of pattern recognition systems and feature extraction systems which is used by most of the authors in their work [53, 60, 61] for measuring the toxicity in food products.

In order to simplify the explanation of how feature extraction and pattern recognition methods can be applied for detecting toxicity in food products, by using the MQ gas sensors in the most simple and economical way, let us take the help of tinkercad to simulate our model of e-nose that we can use for the detection of toxicity in food samples as shown in Figs. 10, 11 and 12.

In our simulation, we interface a toxic gas assumed to be emitted by a contaminated food sample with the e-nose. The data collected during this process represents the response of the MQ gas sensor array to changes in the target environment. This

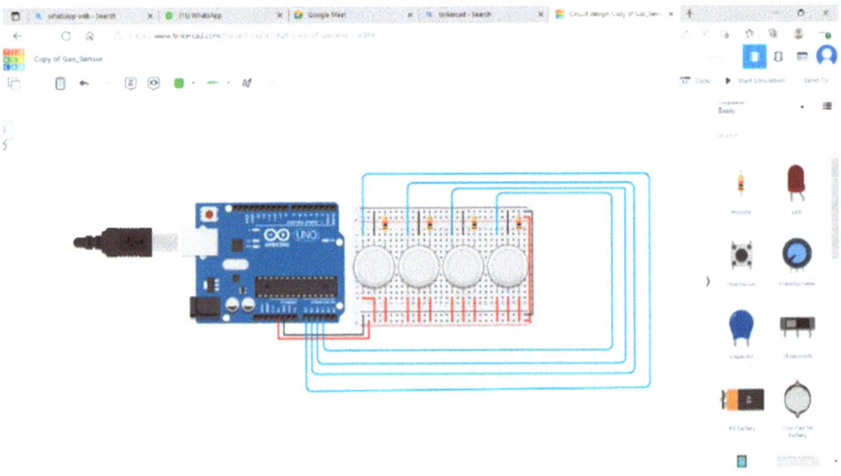

Fig. 10 Screenshot of the e-nose system prepared in tinkercad virtually

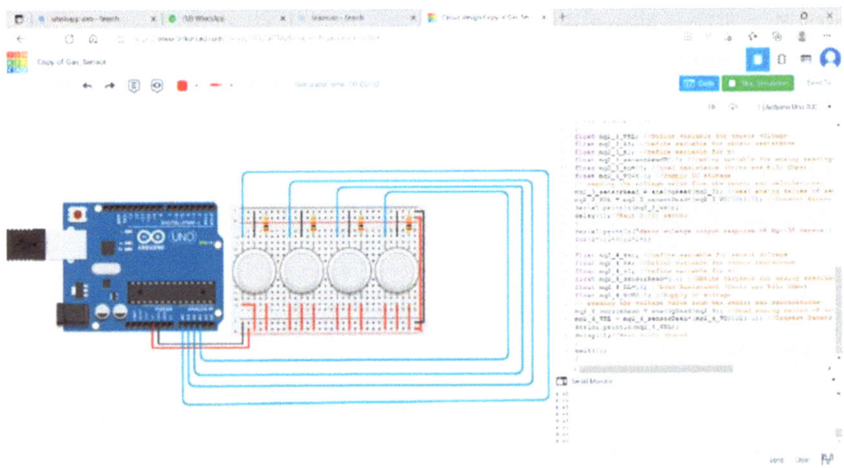

Fig. 11 Screenshot showing running of the simulation for data collection in tinkercad

response is observed as random variations in the voltage drop across the respective sensors. When these data points are plotted shown in Fig. 13, they produce a pattern that reflects the characteristic response of the e-nose, as illustrated below:

Now, the above graph according to our assumption shows the response of the e-nose system against the presence of some toxic that was released by a toxic food sample. We have this response pattern of the e-nose system for a given toxic food sample and we can use it as the basis for detection of toxicity of the same unknown food sample which may or may not be toxic. If the response pattern of the given unknown food sample comes out to be similar to that of the first case which we had

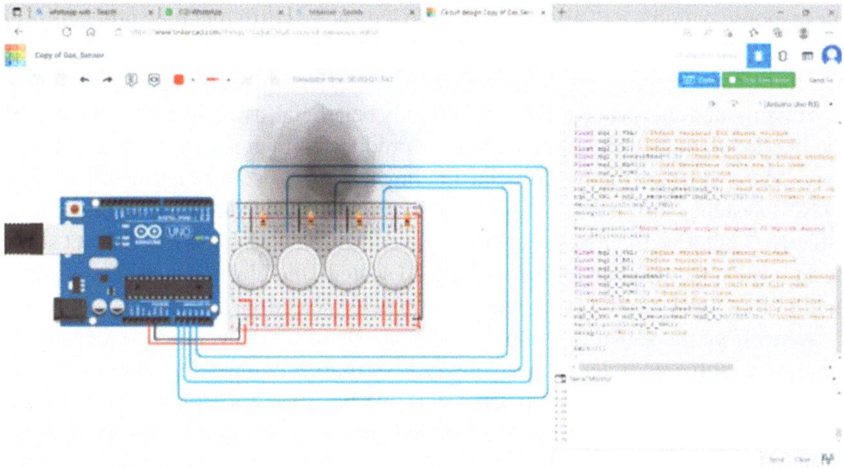

Fig. 12 Screenshot of the tinkercad simulation showing the gas sensors being triggered by simulated smoke

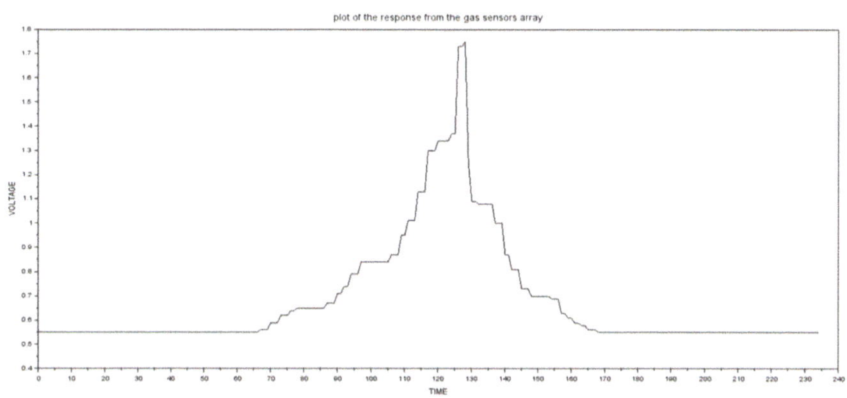

Fig. 13 Response of the MQ-gas sensor array to the toxic gas

obtained when the food sample was toxic, we will conclude that the unknown food sample taken in the second case is also toxic, and however if the response does not match with the one that we had obtained in the first case we will conclude that the given unknown food sample in the second case is not toxic in nature and can be consumed safely. However, even if this process seems to be simple; in reality more précised approach and technical methods needs to be used for better results, which can help in preparing a market competent device. And, for comparing these responses we can make use of artificial intelligence or artificial neural network systems as mentioned in our earlier discussion.

Once the response pattern of the e-nose system for a specific type of toxicity in a particular food product or sample is obtained, we can extend this process to identify the response patterns for various types of toxicity across different food products. This approach allows us to build a comprehensive library of data containing information about response patterns for diverse toxicities in different food samples.

Having such an extensive database is highly beneficial, as a larger dataset increases the likelihood of detecting toxicity in food products with greater precision and accuracy. Moreover, leveraging artificial intelligence (AI) or artificial neural network (ANN) systems for pattern recognition and matching becomes significantly more effective with a well-curated and extensive collection of data. A rich dataset serves as an invaluable resource for training the AI system, thereby enhancing its accuracy and improving its overall performance in toxicity detection.

Apart from the above-mentioned challenges there are many other challenges that we have to solve, in order to make any kind of market competent economic device, which can be useful for the society. Here we have tried to solve only two basic problems, that are needed to be studied and solved in the initial steps itself- firstly the creation of a cost-effective and efficient vacuum chamber and secondly creation of an effective e-nose system. Hence, in near future we will continue our work for a better system trying to solve other complications or issues that comes up for detection of harmful substances in food products and efficient enough to be useful for the society.

4 Conclusion

The development and application of electronic nose (e-nose) technology have emerged as a promising approach to addressing critical challenges in food adulteration detection and quality control. As discussed in this chapter, e-noses, which replicate the human olfactory system using gas sensor arrays and advanced algorithms, provide a fast, sensitive, cost-effective, and reliable alternative to traditional methods. These systems are particularly advantageous in applications where conventional approaches may be slow, expensive, and prone to subjective errors.

While significant progress has been made in enhancing e-nose efficiency, data processing, and classification models, there remain challenges that require further investigation. Issues such as the need for frequent sensor recalibration, the high cost of advanced sensor configurations like SPME-MS-based systems, and the limitations in online monitoring must be addressed to make e-noses more accessible and versatile. Advancements in robust sensor materials, improved algorithms for noise reduction, and machine learning techniques inspired by the olfactory system hold immense potential for overcoming these challenges.

The current applications of e-nose technology in detecting food adulteration, identifying geographical origins, and monitoring spoilage and contamination demonstrate its value in the food industry. However, the dream of a universal e-nose capable of handling all odour types remains unrealised due to the specific instrumentation and data requirements for different applications. Continuous research and development are essential to refine e-nose systems, making them more user-friendly, cost-effective, and capable of delivering accurate results across diverse scenarios.

Looking ahead, the integration of innovative sensor technologies and AI-driven solutions will pave the way for versatile e-nose systems that can contribute meaningfully to society. A reliable, economical, and accessible e-nose setup has the potential to transform food safety practices and address food adulteration issues on a global scale. By focusing on solving existing limitations and developing robust yet cost-efficient systems, we can create a technology that not only serves its purpose but also adds significant value to research, industry, and public health.

References

1. W. Andrew, *Sensors and Transducers* (2019). https://doi.org/10.1017/9781316286210.003
2. M. Roy, B.K. Yadav, Electronic nose for detection of food adulteration: a review J. Food Sci. Technol. 59 (2021). https://doi.org/10.1007/s13197-021-05057
3. A.T. John, K. Murugappan , D.R. Nisbet, A. Tricoli, *An Outlook of Recent Advances in Chemiresistive Sensor-Based Electronic Nose Systems for Food Quality and Environmental Monitoring* (2021)
4. N. Dutta, chapter 1, *Noise Feature Analysis, System Identification and Modelling for Selection of Pulse Temperature Frequency of MOS Gas Sensors*
5. A. Berna, Metal oxide sensors for electronic noses and their application to food analysis. Sensors (Basel, Switzerland). **10**, 3882–3910 (2010). https://doi.org/10.3390/s100403882
6. O. Salih, S. Mohammed, H. Robert, Z. Zejun, H. Lars, K. Adnan, W. Christof, N. Peter, L. Uli (2020) *Identification of Mint Scents Using a QCM Based E-Nose.* 10.20944 preprints202012.0791.v1
7. A. Kanade, A. Shaligram Assistant Professor, Department of Electronic Science, P.V.P. College, Pravaranagar, India1 Professor and Head, Department of Electronic-Science, Review on MOS Gas Sensors based e-Noses and their Applications to Food Analysis
8. D. Tomasz, C. Tomasz, W. Waldemar, Invited ReviewArticle: An odor-sensing system powerful technique for foodstuff studies. Rev. Sci. Instrum. **82**, 1–32 (2011).https://doi.org/10.1063/1.3660805
9. K.U. Diclehan, U. Oguzhan, T. Mehmet, Electronic nose and its applications: a survey. Int. J. Autom. Comput. **1–31**(10), 1007 (2019). s11633–019–1212–9
10. Md Al Mamun, B. Biplob, T. Sumita, Md. Islam, An overview of food adulterants and their health impacts. Int. J. Sci. Res. Publ. (IJSRP). 11. 780–796 (2021). https://doi.org/10.29322/IJSRP.11.05.2021.p11381
11. S. Bartosz, A. Krzysztof, N. Jacek, G. Jacek, Determination of odour interactions in gaseous mixtures using electronic nose methods with artificial neural networks. Sensors (Basel, Switzerland) **18** (2018). https://doi.org/10.3390/s18020519
12. M. Peris, L. Escuder-Gilabert, Electronic noses and tongues to assess food authenticity and adulteration. Trends Food Sci. Technol. **58** (2016). https://doi.org/10.1016/j.tifs.2016.10.014
13. K. Oguzhan, U. Mehmet Turkan Department of Electrical and Electronics Engineering, Electronic Nose and Its Applications: A Survey Diclehan Izmir University of Economics, Izmir 35330, Turkey
14. W. Jia, G. Liang, Z. Jiang, J. Wang, Advances in electronic nose development for application to agricultural products. Food AnalMethods **12**(10), 2226–2240 (2019)
15. Q. Wang, L. Li, W. Ding, D. Zhang, J. Wang, K. Reed, B. Zhang, Adulterant identification in mutton by electronic nose and gas chromatography-mass spectrometer. Food Control **98**, 431–438 (2018)

16. N. Subari, J. Mohamad Saleh, A. Yeon Md Shakaff, *Fusion Technique for Honey Purity Estimation using Artificial Neural Network* (2014)
17. S.K. Pandey, Prevailing practices of artificial ripening of mango, banana and papaya through calcium carbide in Jharkhand. AgricEng Today **40**(3), 35–39 (2016)
18. M. Mir-Hassan, K.K. Houman, K. Amin, H. Sohrabi, Assessment of the chemical adulteration and hygienic quality of raw cow milk in the northwest of Iran. Qual. Assur. Saf. Crop. Foods. 1–8 (2019). https://doi.org/10.3920/QAS2019.1605
19. M. Rasekh, H. Karami, E-nose coupled with an artificial neural network to detection of fraud in pure and industrial fruit juices. Int. J. Food Prop. **24**2021https://doi.org/10.1080/10942912.2021.1908354
20. S. Tatli, E. Mirzaee-Ghaleh, H. Rabbani, H. Karami, A. Wilson, Rapid detection of urea fertilizer effects on VOC emissions from cucumber fruits using a MOS E-nose sensor array. Agronomy **12**, 35 (2021). https://doi.org/10.3390/agronomy12010035
21. S. Mildner-Szkudlarz, H. Jelen, The potential of different techniques for volatile compounds analysis coupled with PCA for the detection of the adulteration of olive oil with hazelnut oil. Food Chem. **110**, 751–761 (2008). https://doi.org/10.1016/j.foodchem.2008.02.053
22. B. Eva, F. Joan, B. Ricard, M. Montse, A. Laura, B. Olga (2015) Data fusion methodologies for food and beverage authentication and quality assessment—a review. Anal. Chim. Acta. **891**.https://doi.org/10.1016/j.aca.2015.04.042
23. P. Boeker, On 'Electronic Nose' methodology. Sens. Actuators, B Chem. **204**, 2–17 (2014). https://doi.org/10.1016/j.snb.2014.07.087
24. Z. Hai, J. Wang, Detection of adulteration in camellia seed oil and sesame oil using an electronic nose. Eur. J. Lipid Sci. Technol. **108**, 116–124 (2006). https://doi.org/10.1002/ejlt.200501224
25. Z. Hai, J. Wang, Electronic nose and data analysis for detection of maize oil adulteration in sesame oil. Sens. Actuators, B Chem. **119**, 449–455 (2006). https://doi.org/10.1016/j.snb.2006.01.001
26. M.C. Cerrato Oliveros, J.L. Pe´rezPavo´n, C. Garcia Pinto, M.E. Ferna´nLaespada, B. Moreno Cordero, M. Forina, Electronicnose based on metal oxide semiconductor sensors as a fast alternative for the detection of adulteration of virgin olive oils. Anal. Chim. Acta 459, 219–228 (2002)
27. M. Bougrini, K. Tahri, Z. Haddi, T. Saidi, N. El Bari, B. Bouchikhi, Detection of adulteration in Argan oil by using anelectronic nose and a voltammetric electronic tongue. J. Sens. **2014**, 1–10 (2014)
28. A. Marina, Y. Man, A. Ismail, Use of the SAW sensor electronic nose for detecting the adulteration of virgin coconut oil with RBD palm kernel Olein. J. Oil & Fat Ind. **87**, 263–270 (2009). https://doi.org/10.1007/s11746-009-1492-2
29. E. Hong, P. Sue, C. Jin, N. Bong, Discrimination of palm olein oil and palm stearin oil mixtures using a mass spectrometry based electronic nose. Food Sci. Biotechnol. **20**, 809–816 (2011)https://doi.org/10.1007/s10068-011-0112-8
30. W. Xiaobao, S. Xingfeng, W. Yingying, C. Ling-Zhi, P. Lei-qing, T. Kang, Rapid detection of adulterated peony seed oil by electronic nose. J. Food Sci. Technol. **55** (2018).https://doi.org/10.1007/s13197-018-3132-z
31. H. Yu, J. Wang, Y. Xu, Identification of adulterated milk using electronic nose. Sensor Mater **19**, 275–285 (2007)
32. T. Mojtaba, G.-V. Mahdi, G. Vahid, M. Seyed, B. Mojtaba, Identification of trace amounts of detergent powder in raw milk using a customized low-cost artificial olfactory system: a novel method. Measurement **124** (2018).https://doi.org/10.1016/j.measurement.2018.04.006
33. F. Ayari, E. Mirzaee-Ghaleh, H. Rabbani, K. Heidarbeigi, Detection of the adulteration in pure cow ghee by electronic nosemethod (case study: sunflower oil and cow body fat). Int. J. Food Prop. **21**(1), 1670–1679 (2018)
34. F. Ayari, E. Mirzaee-Ghaleh, H. Rabbani, K. Heidarbeigi, Usingan E-nose machine for detection the adulteration of margarine in cow ghee. J. Food Process Eng **41**(6), e12806 (2018)
35. M. Jamilah Masnan, A. Zakaria, A. Yeon Md. Shakaff, N. Idayu Mahat, H. Hamid, N. Subari, J. Mohamad Saleh, *Principle Component Analysis—A Realisation of Classification Success in Multi Sensor Data Fusion* (2011)

36. N. Subari, J. Mohamad Saleh, A. Yeon Md Shakaff, A. Zakaria, *A Hybrid Sensing Approach for Pure and Adulterated Honey Classification* (2012)
37. Z. Ammar, Md S. Aliyeon, M. MazJamilah, A. Mohd, A. Abdul, J. Mahmad Nor, S.A. Ghani, A. Abu Hassan, A. Abdul, K. Latifah, S. Norazian, F. Nazifah, A biomimetic sensor for the classification of honeys of different floral origin and the detection of adulteration. Sensors (Basel, Switzerland). **11**, 7799–822 (2011). https://doi.org/10.3390/s110807799
38. Q. Peng, R. Tian, F. Chen, B. Li, H. Gao, Discrimination of producing area of Chinese Tongshan kaoliang spirit usingelectronic nose sensing characteristics combined with The chemometrics methods. Food Chem. **178**, 301–305 (2015)
39. M. Penza, G. Cassano, Recognition of adulteration of Italian wines by thin—film multisensor array and artificial neural networks. Analytica Chimica Acta—ANAL CHIM ACTA. **509**, 159–177 (2004). https://doi.org/10.1016/j.aca.2003.12.026
40. E. Ordukaya, B. Karlik, Fruit juice—alcohol mixture analysis using machine learning and electronic nose. IEEJ Trans. Electr. Electron. Eng. **11**, S171–S176 (2016). https://doi.org/10.1002/tee.22250
41. X. Hong, J. Wang, Detection of adulteration in cherry tomato juices based on electronic nose and tongue: comparison of different data fusion approaches. J. Food Eng. **126**, 89–97 (2014)
42. X. Hong, J. Wang, S. Qiu, Authenticating cherry tomato juices discussion of different data standardization and fusion approaches based on electronic nose and tongue. Food Res. Int. **60**, 173–179 (2014)
43. H. Wang, C. Xueli, H. Tian, P. Hairun, R. Hong, S. Sara, A novel methodology for real-time identification of the botanical origins and adulteration ofhoney by rapid evaporative ionization mass spectrometry. Food Control **106**, 106753 (2019). https://doi.org/10.1016/j.foodcont.2019.106753
44. U. Banach, C. Tiebe, T. Hübert, Multi gas sensors for the quality control of spice mixtures. Food Control **26**, 23–27 (2012)
45. K. Heidarbeigi, S.S. Mohtasebi, A. Foroughirad, Ghasemi-Varnamkhasti adulteration in saffron samples using electronic nose. Int. J. Food. Prop. **8**, 1391–1401
46. S. Kiani, S. Minaee, M. Ghasemi-Varnamkhasti, Integration of computer vision and electronic nose as non-destructive systems for saffron adulteration detection. Comput. Electron. Agric. 141 (2017)https://doi.org/10.1016/j.compag.2017.06.018
47. K. Rahul, J. Jainendra, R. Pallavi (2016) Therapeutic potentials of cow derived products-a review. Int. J. Pharm. Sci. Res. **7**, 1383–1390. https://doi.org/10.13040/IJPSR.0975-8232.7(4).1383-90
48. K. Gandhi, A. Kumar, D. Lal, Solvent fractionation technique paired with apparent solidification time (AST) test as a method to detect palm olein and sheep body fat in ghee (clarified milk fat). Indian J. Dairy Science. **71**, 246–251 (2018)
49. N. Subhash Yadav, S. Jyoti Prakash, S. Kailash, S. Smrutilekha, *Detection of Adulteration in Ghee—A Spoonful of Yellow Magic* **2**, 912–914 (2020). https://doi.org/10.13140/RG.2.2.14907.95524
50. A.L. Amrutha Kala, *Detection of Possible Adulteration in Commercial Ghee Samples Using Low-Resolution Gas Chromatography Triglyceride Profiles* (2013). https://doi.org/10.1111/1471-0307.12049
51. F. Korel, M.Ö. Balaban, Quality assessment of aquatic foods by machine vision. Electron. Nose, Electron. Tongue (2010). https://doi.org/10.1002/9781444325546.ch6
52. M. Roy, N. Hariharan, D. Manoj, S. Shanmugasundaram, Development of metal oxide semiconductor gas sensor based electronic nose system for adulteration detection in ghee
53. U. Banach, C. Tiebe, T. Hübert, Multi gas sensors for the quality control of spice mixtures. Food Control **26**, 23–27 (2012)
54. Y.B. Man, H.L. Gan, I. NorAini, H. Nazimah, T. Chin, Detection of lard adulteration in RBD palm olein using an electronic nose. Food Chem. **90**, 829–835 (2005). https://doi.org/10.1016/j.foodchem.2004.05.062
55. Z. Gan, Y. Yang, J. Li, X. Wen, M. Zhu, Y. Jiang, Y. Ni, Usingsensor and spectral analysis to classify botanical origin anddetermine adulteration of raw honey. J. Food Eng. **178**, 151–158 (2016)

56. R. Garrido-Delgado, Ma. Muñoz-Pérez, L. Arce, Detection of adulteration in extra virgin olive oils by using UV-IMS and chemometric analysis. Food Control. 85 (2017)https://doi.org/10.1016/j.foodcont.2017.10.012
57. F. Tsopelas, D. Konstantopoulos, A. Tsantili Kakoulidou, *Voltammetric Fingerprinting of Oils and its Combination with Chemometrics for the Detection of Extra Virgin Olive Oil Adulteration* (2018)
58. M.C. Cerrato Oliveros, J.L. PeŕezPavoń, C. Garcia Pinto, M.E. Fernań Laespada, B. Moreno Cordero, M. Forina (2002) Electronic-nose based on metal oxide semiconductor sensors as a fast alternative for the detection of adulteration of virgin olive oils. Anal. Chim. Acta. **459**, 219 228
59. D. Abigail, *Food Adulteration* (2022). https://doi.org/10.1007/978-3-030-02721-6_292-1
60. J. Yamawaku et al., Vacuum chamber fast dehumidification ~water control in semiconductor manufacturing equipment. Int. Symp. Semicond. Manuf. **2007**, 1–4 (2007). https://doi.org/10.1109/ISSM.2007.4446884
61. V. Basavarja , L. Murali Kumar, G.N. Sahadeva, Design and analysis of vacuum chamber cover. Int. J. Eng. Res. & Technol. (IJERT) **07**(07) (2018)
62. M. Billing et al., Measurements of vacuum chamber impedance effects on the stored beam at CESR, in *Proceedings Particle Accelerator Conference*, vol. 5 (1995), pp. 3206–3208. 10.1109 PAC.1995.505830
63. M. Falasconi, I. Concina, E. Gobbi, V. Sberveglieri, A. Pulvirenti, G. Sberveglieri, quot, Electronic nose for microbiological quality control of food products & quot. Int. J. Electrochem. (2012)

Unveiling the Synergy of Nanocolorimetric and Plasmonic Based Techniques Towards Sensing of Food Adulterants

Upama Das and Nikhil Kumar Daimari

Abstract Globally growing demands of food supply chain has heightened concerns regarding food safety where rising instances of food adulteration has created an urgent need for rapid, cost-effective, and accurate sensing technologies. In recent years nano colorimetric and plasmonic based sensing techniques have been studied extensively as innovative approach for the detection of food adulterants. In particular, colorimetric sensors driven by nanoscale mechanisms, inducing alteration in optical characteristics of the material, have the ability to detect a variety of adulterants. Similarly, plasmonic sensors identifies the same range of contaminants by leveraging surface plasmon resonance for enhanced sensitivity. Here in this chapter, an overview into the principles behind nano colorimetric and plasmonic phenomena have been described highlighting several probes designed for adulterant detection. In addition, the chapter also explores the current challenges in the field and compares the present technique with existing sensing protocols. Lastly, the chapter concludes with a discussion on future perspectives and emerging trends summarizing a comprehensive review on the potential of such methods towards the advancement of food safety monitoring.

Keywords Nanoparticles · Colorimetric · Plasmonic · Food adulteration · Preservatives · Dyes

---

U. Das (✉) · N. K. Daimari (✉)
Applied Optics and Photonics Research Laboratory, Department of Physics, Tezpur University, Tezpur, India
e-mail: upamads@tezu.ernet.in

N. K. Daimari
e-mail: nico_21@tezu.ernet.in

R. Biswas (ed.), *Optical Techniques for Assessing Food Adulterants*, Progress in Optical Science and Photonics 35, https://doi.org/10.1007/978-981-96-8142-6_10

1 Introduction

Food being the primary component for survival, growth and overall wellbeing of all human life. Consumption of high-quality food is crucial for the absorption of necessary nutrients required for the growth and development of human body. Hence, it has become mandatory to check the food quality in order to guarantee the safety of global consumers. Nowadays, various food items are prone to adulteration and fraud due to increasing demand and globalisation of the supply chain. Adulteration of food is considered as an act of intentional contamination, addition of inferior substances and improper use of practices during the processing of any food item. The ulterior motive behind such fraud is mostly related to economic gain, to extend shelf life at low cost and enhancing the value by malpractices such as dilution, substitution etc. leading to compromised nutritional intake, safety and increased health risks. The problem of food adulteration is not new and has been there since beginning of food processing and production systems. Almost all kinds of food products are subjected to food adulteration which includes dairy products, oils, spices, seafood, drinks and even vegetables and fruits that are being sold in the market. Lack of hygiene or ignorance towards proper facilities may also lead to unintentional adulteration of food [1–5].

Numerous health risks are also associated with the consumption of adulterated food items with diseases ranging from mild to life-threatening. The common health hazards are nausea, vomiting, diarrhoea, allergies etc. In addition to that there are also long-term effects on consumption of adulterated food such as heart diseases, kidney problems, ulcers to more extreme diseases as asthma, nerve damage and cancer. One of the international events that have led to major consequences is the milk scandal in China in 2008 where melamine contaminated milk affected around 300,000 infants leading to 6 fatalities and 860 hospitalized [6]. Other scandals such as Irish pork crisis (2008) [7], European horse meat scandal (2013) [8] and cumin scandal in Europe and US (2015) [9] also demonstrates the negative effects of adulteration.

Due to increasing fraudulent activities concerning food items, fabrication of quick, reliable and cost-effective detection methods for identification of adulterants have become an important research agenda in the past few decades. Traditional analytical techniques such as use of high-performance liquid chromatography (HPLC), mass spectrometry (MS), gas chromatography (GC), ion chromatography (IC), including various chemical sensing and use of microbiological assays for food adulterant detection produce results which are highly accurate. But these techniques face multiple challenges due to need of extensive sample preparation, pretreatment and purification which increases the total time required for sensing. The complexity of such techniques is not conducive to large scale applications and hinders the efficient analysis of the samples. Hence innovative and advanced ways of real time monitoring of food is essential to prevent potential risks associated with their adulteration [10, 11].

Over the decade, colorimetric and plasmonic based techniques have emerged as potential detection methods for sensing adulterants and contaminants in food. These kind of sensing platforms are developed utilizing plasmonic materials where plasmon resonance principle is responsible for sensing mechanism. For metallic nanostructures of size much smaller than the wavelength of incident electromagnetic field penetrates into the metal causing the oscillation of the free conduction band electrons coherently in response to the electric field resulting in electric dipole oscillation to have a resonance with that of incident light. These plasmon excitations are localized in nature and the principle is termed as localized surface plasmon resonance (LSPR) (Fig. 1). Thus, by tuning the shape and size of the nanoscale materials or by changing the surrounding media thereby altering the refractive index of the metal surface, it is possible to optimise the optical properties to enhance the performance required for specific applications (Fig. 2). Such detection methods are reliable and offers rapid sensing of analytes and have the potential to be utilized as tools to investigate food adulteration by both government and industrial laboratories. This chapter explores the principles of colorimetric and plasmonic sensors, along with their applications in detecting food adulteration [12–14].

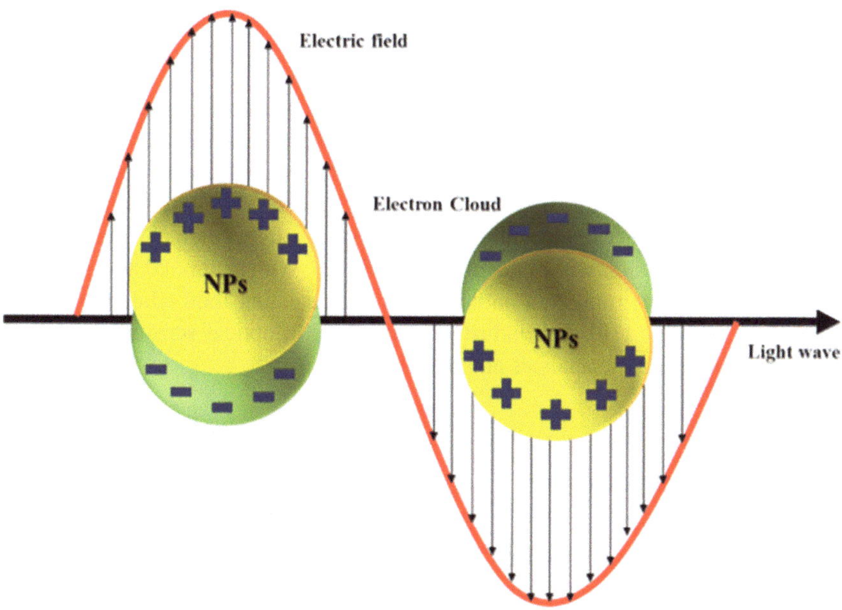

Fig. 1 Pictorial illustration of LSPR displayed by plasmonic nanoparticles

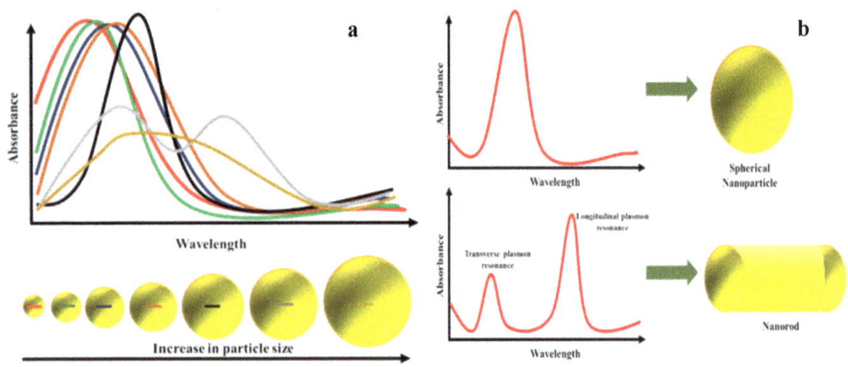

Fig. 2 **a** Graphical representation displaying shift in absorbance peak position with increase in nanoparticles size (Adapted from Ref. [15]. Available under a CC-BY 4.0 license. Copyright © 2023 The Authors), and **b** graphical representation displaying dependence of absorbance with the shape of nanoparticles

2 Principles of Visual Optical Sensing

Optical sensing is a specialized technique used for harnessing photons to glean insights about specific target analytes within a sample. It entails capturing data from light interactions with detectors to discern the presence of undesired substances. Rooted in the fundamentals of optics and photonics, this technology scrutinizes how light behaves when interacting with particular targets, gauging their absorbance, scattering, and transmission within the medium [15, 16]. Widely applied across diverse realms including environmental monitoring [17], detecting food adulteration [18], biomedical imaging [19], and industrial sensing [20], optical sensing serves as a versatile tool for precise analysis and detection.

2.1 Colorimetric Sensors

They are analytical tools that detect the presence of specific substances in samples by monitoring changes in colour. These sensors rely on chemical reactions occurring between the sensor and the target analyte, resulting in discernible alterations in the colour of the sensing material. These colour changes act as visual cues, providing qualitative estimates of the analyte's concentration. The functionality of colorimetric sensors is contingent upon selective chemical interactions and bonding events between the NPs and the target analyte. These interactions induce modifications in the optical characteristics of the sensing nanomaterial, culminating in observable shifts in colour. To validate and quantify these alterations, precise instruments such as colorimeters or spectrophotometers are utilized to accurately measure the intensity of the colour change. In essence, colorimetric sensors exploit chemical reactions to generate distinctive colour changes, offering a straightforward yet efficacious means to detect and estimate the presence of specific analytes in diverse samples [21, 22].

Among the plethora of sensing elements available, nanoscale probes emerge as particularly prominent in contemporary applications. These probes leverage the unique optical characteristics of NPs to discern the presence of specific substances in samples. A wide array of nanoparticle categories can be harnessed, each selected based on their distinct properties [23].

Notably, noble metal nanoparticles are preferred due to their extraordinary plasmonic characteristics occurring within the visible region. Through meticulous functionalization with appropriate ligands or receptors, these nanoparticles exhibit a propensity for highly selective bonding with the target material. The interaction between functionalized nanoparticles and the target material induces alteration in the nanoparticles' characteristics, leading to aggregation or agglomeration phenomena that significantly alter their optical characteristics. This reduction in interparticle distance results in colorimetric changes in nanoparticles, which can be visually observed or confirmed using standard spectrophotometric techniques. The resulting colour change is depending on the amount of the analyte present in the sample [24].

This sensing modality boasts a higher recovery rate and rapid sensing advantages, coupled with lower limits of detection (LOD) and limits of quantification (LOQ), augmented selectivity and sensitivity, towards the detection of the targeted analyte in the sample. This versatile sensing technique finds wide-ranging applications, particularly in food safety monitoring, environmental pollution detection, assessment of pesticide content in food, and identification of food contaminants. Overall, this approach offers a potent strategy for addressing the issue of food adulteration, furnishing real-time assessments of food samples to ascertain the presence of toxic and harmful adulterants in a cost-effective manner [25].

Several research groups have directed their efforts towards utilizing noble metal nanoparticles, such as silver (AgNPs) and gold (AuNPs), for colorimetric detection of food adulterants. Within the realm of colorimetric sensing, two main categories emerge: interference-based colorimetric sensing and rapid colorimetric sensing. In interference-based colorimetric sensing, the change in colour occurs due to the inhibition of formation of nanoparticles in the presence of the targeted analyte. Conversely, in rapid colorimetric sensing, the colour change results from alterations in the interparticle distance within nanoparticles caused by bonding between the targeted analytes and the nanoparticles.

2.1.1 Colorimetric Detection of Artificial Dyes

A broad spectrum of artificial dyes exists, with Sudan dyes representing one such category. These dyes, further segmented into various class of Sudan I, II, III, and IV, dyes which are also classified as lipophilic azo dyes predominantly employed in diverse industrial applications. Notably, their usage in food substances is strictly prohibited due to their high carcinogenic potential. Despite regulatory restrictions, the affordability of Sudan dyes has led to their clandestine incorporation into various food products, including Indian spices like chili powders, as well as sweets, frozen meats, and certain beverages. Consumption of these dyed products poses severe health risks due to their toxic nature. To uphold public health and ensure food safety, it is highly important to tightly regulate the utilisation of Sudan dyes in food items. Effective monitoring and control of these substances can be achieved through the implementation of advanced sensing techniques [26, 27].

To identify Sudan dyes in food items, Wu et al. developed a light scattering technique. The unique molecular composition of these dyes facilitated their role as reducing agents, prompting the formation of AgNPs from silver nitrate ($AgNO_3$). This process was validated by noticeable color transitions from red to brown, indicative of silver nanoparticle generation. A robust plasmonic peak emerged at approximately 452 nm. Limit of determination ranged from 3.0 to 3.2 nM for Sudan I, II, and III, and 2.9 nM for Sudan IV. Subsequent testing on hot chili samples demonstrated recovery rates between 90.8 and 103.3%, with a relative standard deviation of 4.0–4.9% [28].

2.1.2 Colorimetric Detection of Food Preservatives

A broad array of preservatives is intentionally incorporated into food products despite the inherent risks which is directly related to their consumption. These artificial additives are used to inhibit microbial growth, thereby increasing the shelf life of food items; for instance, formalin is utilized in fish and meat to achieve this objective. Nitrites, another commonly used preservative, serve multiple purposes such as enhancing colour, flavour, and texture in processed meats while also acting as inhibitors against the growth of clostridium botulinum spores. However, sodium nitrite, in particular, poses notable health risks. It acts as a precursor to carcinogenic N-nitrosamines, even at minimal concentrations, thereby posing severe health implications [29].

Despite regulatory efforts, the indiscriminate utilization of toxic preservative persists within the food industry, often as a cost-saving measure to replace permitted additives. Consequently, the detection of nitrite in food becomes crucial, particularly in ensuring food quality control and mitigating health risks linked to its consumption. Additionally, various other preservatives such as formaldehyde, salicylic acid, and hydrogen peroxide are commonly added to milk and its derivatives. The consumption of these items poses significant carcinogenic risks and can lead to severe health complications [30].

For detection of nitrite in meat products, Li et al. developed paper-based strips. Their fabricated hybrid gold nanorods-azo-GNPs assemblies, for naked-eye detection of nitrite. This was evident from the colour change to purple from orange-yellow in presence of nitrite, which was found to be highly selective and sensitive for onsite sensing of nitrite [31].

On the other hand, for colorimetric sensing of nitrite in sausage, Xiong et al. developed polyethylene glycol-based AuNPs probe. This probe was designed by functionalising AuNPs using PEGs and recognition ligands. PEGs ensured colloidal stability against real sample conditions, while the other ligand 4-aminobenzenethiol (4-ABT) was used for nitrite sensing. This fabricated sensor was found to be highly stable, selective, and sensitive towards detection of nitrite in complex samples [32].

A novel colorimetric approach was developed by Mathawccsansurn et al. for the sensitive and selective detection of formaldehyde, based on the anti-aggregation property of AuNPs in the presence of melamine. Citrate-capped AuNPs were synthesized using the Turkevich method. Upon the introduction of a specific concentration of melamine, a distinct color change of the AuNPs solution from red to purple occurred, accompanied by a shift in the LSPR peak shifted to 640 nm from 520. But in the presence of formaldehyde, a reaction ensued between formaldehyde and melamine, which impeded the aggregation of AuNPs. A calibration curve was constructed correlating the absorbance ratio (A_{640}/A_{520}) to formaldehyde concentration within the range of 0–2000 $\mu g\ L^{-1}$. The method demonstrated a LOD of 26.0 $\mu g\ L^{-1}$ and a LOQ of 88.0 $\mu g\ L^{-1}$. Moreover, it exhibited high recovery between 97.3 and 104% and precision [33].

Agharkar et al. introduced an innovative method for detecting formaldehyde in milk samples. The formalin in milk interacted with AuNPs suspended in Tollen's reagent, this interaction led to colour change [34].

In a research work performed by Teodoro et al., a straightforward method was developed to fabricate AgNPs functionalised with a polysaccharide which is cellulose nanowhiskers, which functioned as a colorimetric probe for hydrogen peroxide (H_2O_2) detection. The sensing is based on the phenomenoen of catalytic decomposition of AgNPs, leading to a reduction in the absorption peak at 410 nm proportionate to the concentration of H_2O_2. This reduction exhibited linear correlation with H_2O_2 concentration within ranges between 60–600 μM and 0.01–30 μM, with LOD value of 112 μM and 0.014 μM, respectively [35].

2.1.3 Colorimetric Detection of Drug Residues

Worldwide regulations have been implemented to oversee and mitigate the occurrence of drug residues in various food items, notably in products e.g., milk, meat, eggs and fish. These regulations are essential for safeguarding public health and addressing concerns related to antibiotic resistance in humans, which can arise from the consumption of food containing drug residues. Traditional methods employed for sensing drug residues typically involve sophisticated techniques. While effective, these methods are characterized by their labor-intensive nature and time-consuming processes. Despite their accuracy, the complexity and resource-intensive requirements of these techniques underscore the need for alternative, more streamlined approaches to ensure food safety and regulatory compliance [36].

The group Kwon et al., developed a straightforward method for colorimetric sensing of penicillin which is an antibiotic used in pork meat, using antibody-functionalized platinum nanoparticles (Ab-Pt) along with hybrid magnetic nanoparticle clusters (HMNCs). The amine-functionalized HMNCs, captures the antibiotics and Ab-Pt NPs binds to the antibiotic-HMNC complexes, resulting in the formation of Pt-antibiotic-HMNCs complex, which is then added to 3,3′,5,5′-tetramethylbenzidine (TMB) solution, this induces a colour change [37].

Gukowsky et al., developed a method for detecting gentamicin, which is class of antibiotics using cysteamine (cys)-modified-AuNPs. The limit of detetion of cys-AuNPs was found to be 12.45 nM in water and 100 nM in skimmed milk [38].

2.1.4 Colorimetric Detections of Illegal Food Adulterants

Different categories of adulterants are incorporated into food items to maximize profits, yet these additives render the food unsuitable for consumption due to various associated health risks. Among these adulterants are melamine and urea, which are intentionally added to milk to falsely indicate higher protein content. These compounds, abundant in nitrogen, falsely elevate the perceived protein levels in the milk.

In their study, Das et al. synthesized plasmonic AgNPs by using apple peel extract, serving as a functionalizing as well as a reducing agent. These NPs were used as a colorimetric probe to detect melamine in milk, which induced aggregation in NPs, leading to a visible change in colour. This detection method was validated through UV–Vis spectroscopy, showing a linear decrease in absorbance at 423 nm as the concentration of melamine increased, with a lowest limit of sensing of 0.5 ppm [39]. Similarly, the same group used green tea encapsulated AgNPs for sensing melamine in milk with a LOD of 1.44 ppm (Fig. 3) [40].

A study conducted by Dutta et al. showcased urea sensing in milk using AgNPs, by using both uncapped and green tea functionalized AgNPs. Interaction of the nanoparticles with urea induced a colorimetric change. The developed model accurately measured urea concentration in milk ranging from 6 to 90 mg/dL [41].

Chaisiwamongkhol et al., developed a colorimetric probe for sensing Sibutramine, a previously utilized appetite suppressant in food supplements that has since been banned due to its severe side effects. Their method involves a straightforward and onsite sensing setup employing AuNPs based smartphone system for analysis of sibutramine. The presence of sibutramine causes the aggregation of functionalised AuNPs, resulting in a color change, which can also be monitored via smartphone. By analysing smartphone-captured photo, a linear correlation in the range of 5–15 μM was obtained. The LOQ and LOD were determined as 3.47 μM and 1.15 μM, respectively. The accuracy of sibutramine detection fell within the range of 92–107% [42].

In a study led by Ali et al., pork adulteration in meatball preparations of beef and chicken was determined utilizing AuNPs. The AuNPs undergo a change in

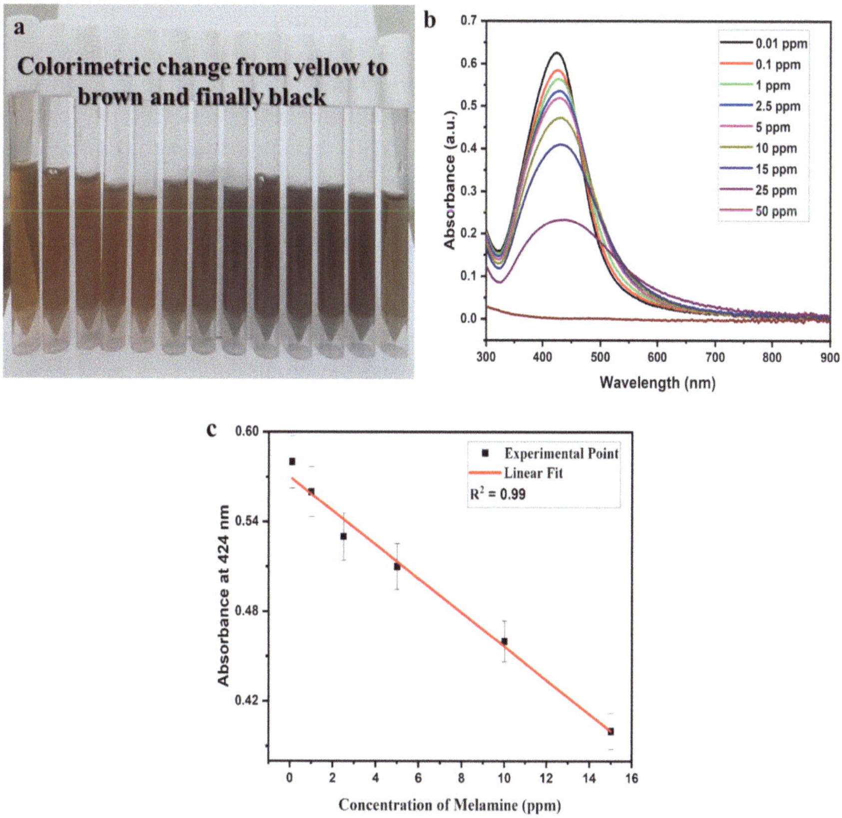

Fig. 3 **a** Colorimetric change in AgNPs in response to melamine adulterated milk supernatant, **b** UV–Vis spectra corresponding to the colorimetric change, and **c** Linear calibration graph between absorbance and increasing concentration of melamine in milk supernatant (Adapted from Ref. [40]. Available under a CC-BY 4.0 license. Copyright © 2024 The Authors)

colour to gray-purple, which shifts their LSPR peak by 30–50 nm. In case of DNA-containing vials, of all non-pork samples remain available and it interacts with AuNPs to prevent salt-induced aggregation. Conversely, meatballs adulterated with pork meat, probes were completely consumed through perfect hybridization, leading to a gray coloration indicating aggregation. This distinction is evident through a distinct red-shift of the absorption peak [43].

Bittar et al. devised a sensitive technique for detecting presence of lead in honey samples, employing stabilizer-free AgNPs. The AgNPs were synthesized via chemical route using borohydride reduction method. Spectrophotometric analysis at 393 nm revealed a linear sensing range spanning from 0.0500 to 0.167 mg L^{-1}, with LOQ and LOD of 0.0451 and 0.0135 mg L^{-1} and, respectively [44].

2.1.5 Colorimetric Detection of Pesticides Residues

Pesticides are indispensable in modern agriculture for pest control and achieving higher crop yields. Nevertheless, the accumulation of residues of pesticide in food products poses significant concerns due to their potential harm to the environment and also on the living organisms. Residues persist in food even after harvesting and processing, accumulating in various items such as fruits, vegetables, grains, and animal products, thereby potentially exposing consumers through dietary intake. Their consumption are associated with severe health hazards, including acute poisoning and chronic health risks. Despite regulatory measures, the persistent presence of pesticide residues underscores the ongoing need for consumer awareness, strict adherence to safety standards, and the promotion of sustainable agricultural practices to mitigate pesticide exposure and safeguard public health and environmental integrity [45].

Chen et al. have fabricated citrate-capped AuNPs for highly selective and rapid detection of dimethoate (DMT) and terbuthylazine (TBA) pesticides. The system allows for the visual detection of TBA and DMT with a LOD value of 0.3 μM and 20 nM, respectively. Moreover, the linear relationships observed through UV–Vis spectrometry for TBA was within the range between 0.1 and 0.9 μM and as in the case for DMT, it was within the range between 1 and 40 nM [46].

Liu et al., employed citrate-coated AuNPs as colorimetric probes to detect cartap pesticide in agricultural crops. Detection relied on the colour transformation of the AuNPs due to its aggregation. A low LOD value of 0.04 mg/kg was obtained by them [47].

For detection of organophosphorus pesticides (OPs) Wu et al., fabricated a scheme mixing acetylcholinesterase enzyme with a solution containing AuNPs and Au^{3+}-cetyltrimethylammonium bromide (Au^{3+}-CTAB). In the absence of OPs, the enzyme produces thiocholine, which protects the AuNPs from dissolving in the solution. However, in the presence of OPs, enzyme activity gets decreased, leading to minimal thiocholine production and an excess of Au^{3+} in the solution. This excess Au^{3+} dissolves the AuNPs, causing a noticeable color change. The method can detect OPs at concentrations as low as 0.7 ppb, and a dipstick coated with AuNPs further enhances sensitivity, with an observable LOD of 35 ppb [48].

2.2 *Plasmonic Sensors for Food Adulterants*

LSPR based sensors have earned significant attention owing to their promising sensing abilities which can be used in various platforms such as sensing of food adulteration, offering rapid and sensitive detection methods that are crucial for ensuring food safety and authenticity. Their unique optical property plays a major role which can detect slight alteration in the surrounding environment caused by adulterants or contaminants in food samples. By exploiting phenomena like LSPR, these sensors can accurately sense the minute changes in refractive index or molecular interactions, enabling the identification of adulterants even at trace levels. Furthermore, plasmonic sensors can be integrated together into on-site portable easy to use user-friendly devices, allowing for real-time testing of food adulteration. The recent development and research in this field hold the significant potential to enhance the capabilities of these sensors across widespread application in the field of food adulteration detection [49–51].

2.2.1 Plasmonic Detection of Artificial Dyes

Wu et al. presented a technique for detecting Sudan azo dyes utilizing a visual light scattering method, which relies on the creation of AgNPs. Sudan dyes act as potent reducing agents due to their specific molecular characteristics, facilitating a redox reaction with $AgNO_3$. This reaction triggers the formation of AgNPs, resulting in noticeable color alterations indicating formation of AgNPs. A robust plasmon resonance light scattering (PRLS) signals was identified at 452 nm which depends on the concentrations of each Sudan dye variant within defined ranges [28].

Xu et al. have also developed an alternative approach for detecting these dyes by creating SPR sensor, coupled with a molecularly imprinted polymer (MIP) nanofilm of thickness 75 ± 5 nm. The sensor was constructed by in situ polymerisation process in which gold chip was attached with the Sudan I MIP nanofilm. The sensor demonstrated remarkable adsorption capacity, along with high selectivity and sensitivity. The SPR response exhibited a linear graph for Sudan I within the dynamic range of sensing between 50–400 ng mL^{-1}. A limit of detection of 30 ng mL^{-1} was obtained by them, with a response time of 400 s [52].

2.2.2 Plasmonic Detection of Food Preservatives

Borah et al. devised a straightforward technique for detecting food preservatives like formaldehyde and benzaldehyde. They employed a robust, sensitive, and selective, non-enzymatic route utilising GA-AuNP@Tollens'. The linear range of sensing was between 10 and 150 nM for formaldehyde and 0.15–0.75 μM for benzaldehyde. The detection limits for benzaldehyde and formaldehyde achieved with this method were 0.12 μM and 20.08 nM, respectively [53].

2.2.3 Plasmonic Detection of Illegal Food Additives

Moznuzzaman et al. have introduced an innovative approach for formalin detection, utilizing 2D graphene-$PtSe_2$ nanocomposite combined with AgNPs and ZnO nanolayer. The sensor demonstrates remarkable sensitivity of 155.33° RIU^{-1} (Refractive index unit) and quality factor of 88.89 RIU^{-1}, respectively. Employing attenuated total reflection (ATR), the sensor detects the presence of formalin by analysing various parameters such as reflectance, SPR angle, transmittance along with surface plasmon resonance frequency (SPRF) [54].

Qi et al. detected formalin in octopus and chicken flesh for which they used Au nano prism/Tollens' reagent. The alteration in colour, attributed to the generation of a new complex which is formaldehyde-tollens complex, served as an indicator for the presence of formaldehyde [55].

2.2.4 Plasmonic Detection of Drug Residues in Food

Antibiotics, being compounds that target specific molecules, are conducive to detection using MIP biosensors. To detect vancomycin in skimmed milk samples Altintas et al. crafted a MIP-SPR sensor designed for detecting. A notable LOD of 4.1 ng/mL when employing a AuNPs-conjugated vancomycin assay. These sensors demonstrated heightened sensitivity levels, surpassing the threshold required for antibiotic quantification [56].

Similarly, to detect milk samples containing tetracycline, Gao et al. developed a sensor where they combined magnetic molecularly imprinted polymer nanoparticles (MMIP NPs) and SPR technology. Here, the SPR chip was modified to achieve highly sensitive tetracycline detection [57].

To determine enrofloxacin in various animal-derived food products, such as chicken muscle, milk, eggs, and duck samples Pan et al. devised an SPR immunosensor. This sensor tracked antigen–antibody reactions to detect the presence of enrofloxacin [58]. To detect gentamicin (GEN) and chloramphenicol (CAP) in milk samples, Xia et al. developed an immune inhibition assay. This sensor enabled simultaneous detection of GEN and CAP using an elution strategy [59].

2.2.5 Plasmonic Detection of Illegal Food Adulterants

To determine adulteration of honey by water, Rahad et al. created an innovative refractive index detector based on the phenomenon of surface plasmon polariton, the sensor is made up of silver nanodots in a metal–insulator–metal (MIM) waveguide configuration along with an octagonal cavity resonator. This sensor detects alterations in refractive index by monitoring the alteration in the position of the resonant wavelength. The detector exhibited a maximum sensitivity of 2527.6 nm per RIU [60]. To detect presence of fructose and glucose in adulterated honey samples, Yadav et al. developed a SPR probe utilizing graphene oxide modified silver film coated optical fibre [61].

To determine the contamination of meatballs with cytochrome b genes of horse, Houhoula et al. fabricated AuNPs. The changes in its optical properties were monitored in presence of an acid. The added acid inhibited the agglomeration of AuNPs, resulting in a pink colour for adulterated samples, whereas unadulterated samples exhibited a purple colour with a peak wavelength absorption at 524 nm [62].

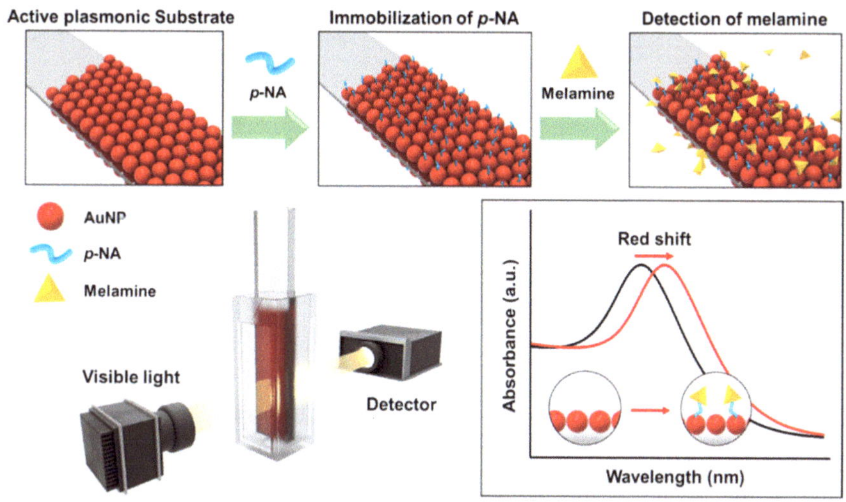

Fig. 4 Pictorial illustration of the cuvette based LSPR chip functionalised with p-nitroaniline-AgNPs immobilised on the glass substrate (Reproduced from Ref. [64]. Available under a CC-BY 4.0 license. Copyright © 2019 The Authors)

For detection of melamine, Li et al. devised a MIP film [63]. Similarly, Oh et al. introduced a cuvette-based LSPR sensor utilizing p-Nitroaniline functionalized AuNPs for detecting melamine in infant formula. This sensor, noted for its simplicity, offers a rapid detection method for assessing adulteration levels in consumer samples (Figs. 4 and 5) [64]. Luna-Moreno et al. utilized SPR based technology to detect adulteration of tequila by methanol. By analyzing the resonant peak, they inferred the ethyl alcohol content, making SPR a valuable technique for differentiating between adulterated and aged tequila [65].

2.2.6 Plasmonic Detection of Pesticides Residues in Food Items

For detection of profenofos, Shrivastav et al., used SPR and optical fiber technology. The sensor displayed a dynamic range of sensing between 10^{-4} and 10^{-1} μg/L [66]. Saylan et al., developed MIP nanofilms to detect pesticides like cyanazine (SNZ), simazine (SMZ), and atrazine (ATZ). They made these nanofilms using UV light and some specific chemicals. These nanofilms were then placed on gold surfaces. They used a device called an SPR sensor to measure these pesticides in real-time. This sensor could detect very small amounts of these pesticides, ranging from 0.10 to 6.64 nM [67].

Yilmaz et al. developed molecularly imprinted nanoparticles to detect pesticides. In this study, they utilized an affinity sensor based on molecular imprinting, where nanoparticles imprinted with atrazine were affixed onto the gold surface. The change

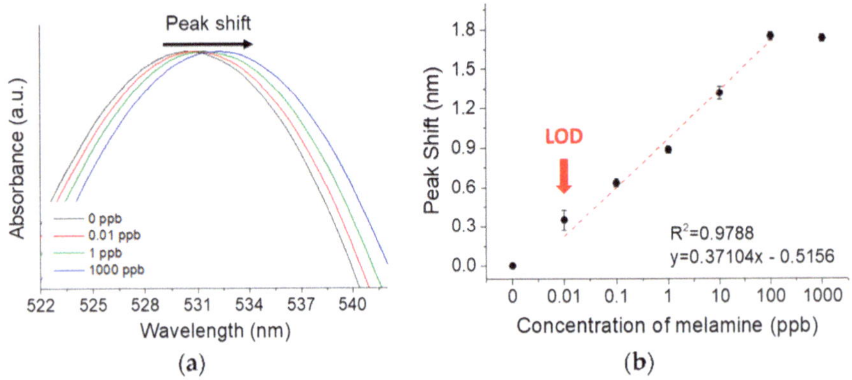

Fig. 5 **a** Detection of melamine by p-nitroaniline functionalised AgNPs based cuvette-type LSPR chip displaying significant shift in plasmonic peak with increasing concentration of melamine in milk, and **b** the linear calibration graph between peak shift and concentration of melamine (Reproduced from Ref. [64]. Available under a CC-BY 4.0 license. Copyright © 2019 The Authors)

in response (ΔR) of the nano sensor to 1.0 ng/mL of atrazine was found to be 0.637 with higher selectivity compared to simazine and amitrole [68, 69].

In another study by Sadrolhosseini et al., fabricated a special layer called polypyrrole-chitosan/graphene oxide (PPy-Chi/GO) to measure napropamide dissolved in hexane, ranging from 0.01 to 100 ppm with an LOD value of 0.1 ppm. They set up the SPR system using a special configuration called Kretschmann configuration. They made layers of polypyrrole-chitosan/graphene oxide with thickness ranging from 8 to 60 nm using a process called electro-deposition. They found that the PPy-Chi/GO layer was able to sense napropamide when they measured the change in angle to be between 0.012° to 2.003°. They could detect napropamide even in very small amounts, with a LOD of about 0.1 ppm [70].

In a study by Miyake et al., they used a special sensor called an SPR immunosensor to identify six different pesticides in samples, such as chlorfenapyr, imazalil, azoxystrobin, boscalid, nitenpyram, and isoxathion. They used a type of sensor chip that had a thin layer of Au on the surface, and this layer was treated with a specific chemical called alkanethiol. This chemical had a carboxy group that was changed to a form called *N*-hydroxysuccinimide. They prepared solutions containing these pesticides individually and also mixed together, along with samples from tomatoes. The highest amount of each pesticide allowed in tomatoes, known as the maximum residue limit, was reported as 1 mg/kg for nitenpyram, 0.01 mg/kg for isoxathion, 3 mg/kg for azoxystrobin, 1 mg/kg for chlorfenapyr, 5 mg/kg for boscalid, and 0.5 mg/kg for imazalil [71].

3 Comparative Analysis and Challenges with Other Sensing Protocols

Optical-based sensing schemes, including colorimetric and plasmonic techniques, offer distinct advantages and challenges compared to other sensing schemes like electrochemical methods. Colorimetric sensors rely on changes in colour to detect analytes, making them simple, cost-effective, and suitable for rapid screening. They are particularly advantageous for on-site testing and visual inspection, making them accessible even in resource-limited settings. Plasmonic sensors, on the other hand, exploit the interaction between light and metallic nanostructures to detect analytes, offering high sensitivity and label-free detection. They excel in detecting small molecules and biomolecules with high specificity, making them valuable in biomedical and environmental monitoring applications. However, optical-based sensing schemes also have some few limitations. Colorimetric sensors may lack quantitative accuracy and sensitivity, especially for low-concentration analytes, and they may be susceptible to interference from background colours or turbidity. Plasmonic sensors, while being highly sensitive, require sophisticated instrumentation in controlled environments, limiting their portability and accessibility. Additionally, both colorimetric and plasmonic sensors may face challenges in multiplexing and real-time monitoring. In contrast, electrochemical sensing schemes, such as voltammetry and amperometry, offer several advantages. They provide quantitative and real-time measurements with high sensitivity and selectivity. Electrochemical sensors can operate in various sample matrices and offer excellent compatibility with miniaturization and integration into portable devices. They are well-suited for continuous monitoring and can detect a wide range of analytes, including biomolecules and heavy metals. However, electrochemical sensors also have drawbacks. They may require complex instrumentation and skilled operators, limiting their widespread adoption in point-of-care testing. Sample preparation and electrode fouling can also affect their performance, and they may be susceptible to interference from redox-active species in the sample matrix. In summary, while optical-based sensing schemes offer simplicity, accessibility, and high sensitivity, electrochemical methods provide quantitative accuracy, real-time monitoring, and versatility. Each technique for detection is chosen based upon the specific need of the application, including selectivity, sensitivity, cost-effectiveness, and portability. Integration of both optical and electrochemical approaches may offer complementary benefits and address the limitations of individual sensing schemes for enhanced analytical performance [72–74].

Optical sensing schemes offer several advantages compared to other sensing techniques in food adulteration detection. Firstly, they are often simpler to use and interpret, making them suitable for rapid screening of food samples without the requirement for complex equipment or extensive sample preparation. Methods such as colorimetric assays provide quick results, facilitating efficient detection of adulterants. Additionally, optical sensing methods tend to be more cost-effective, requiring minimal reagents and instrumentation. This affordability makes them accessible for routine testing in various settings, including resource-limited environments. Moreover, optical sensors can be designed to be portable and handheld, enabling on-site testing and field applications for timely detection of adulterants at the point of production or distribution. Furthermore, these techniques are non-destructive and non-invasive, preserving the integrity of the original food sample for further analysis or consumption. Advanced optical sensing technologies, such as plasmonic sensors and surface-enhanced Raman spectroscopy, offer high sensitivity and selectivity for detecting trace levels of adulterants, ensuring accurate and reliable detection. Optical platforms also have the capability for multiplexing, allowing for the simultaneous detection of multiple analytes within the same sample, increasing throughput and efficiency in food adulteration testing. Overall, optical sensing schemes present a versatile and effective approach for detecting food adulteration, offering advantages in simplicity, affordability, portability, sensitivity, and multiplexing capability compared to other sensing techniques [75–77].

4 Future Perspectives and Trends

As food production becomes more globalised, the risk of adulteration also increases making food safety a critical concern. As a result, there is an escalating need on both regulatory and food producers to adopt faster and efficient methods for thorough monitoring of food. The colorimetric and plasmonic based sensing schemes described in this chapter demonstrated successful detection and quantification of analytes, primarily through non-invasive, cost effective and efficient approaches ensuring the quality and safety of food products. Although significant advancements have been made, there are still numerous aspects that require further exploration in future research. From the standpoint of material fabrication there is significant potential in exploring various noble, non-noble, bimetallic nanomaterials, chiral plasmonic materials [78] alongside photonic nanostructures [79] and conductive polymers [80] which could offer more enhancement in sensor sensitivity and selectivity.

As an emerging technology, potential realization of these kind of sensors spans around a wide range of topics such as integration in digital technology, environmental influence and societal impacts. Simultaneously, digital technologies like Internet of Things (IoT) [81] and integration of AI driven data analytics [82] can help transform how data is sampled, collected and processed. This kind of fusion of digital innovation and nano science paves the way for sensor systems that can provide real time monitoring, greatly enhancing the food quality. Advanced image processing algorithms can further enhance sensor images enabling the extraction of accurate colorimetric data.

Multiplex sensing can also pave an important research direction in this field by incorporating multiplex capabilities in plasmonic and colorimetric sensors which will provide a highly adaptable and flexible solution towards food monitoring. Multi-target sensors will also provide a new approach for monitoring complex samples particularly in food products. Furthermore, utilizing indirect colorimetric sensing for sensor development could expand colour range beyond the traditional colour spectrum while another strategy involves using different fluorescent colours to detect multiple adulterant targets simultaneously [83].

Moreover, applying these sensing techniques into real world on-site challenges can have significant societal benefits particularly in the areas of health and sustainability. Sensors of these kinds, owing to their fast and precise identification of contaminants and pathogens plays a vital role in safeguarding against foodborne illnesses and outbreaks. Additionally, advancement towards eco-friendly and biodegradable sensor materials can also help minimise the effects of waste and promote sustainable practices. Therefore, it is crucial to strike a balance between innovation, social responsibility and environment to create a positive impact in the society.

Lastly, one major area for future development is in the commercialization of these food adulterant sensing platforms. To meet the industrial standards, sensing devices should have some necessary requirements which include stability, reproducibility and cost effectiveness. Innovative techniques and strategies can be employed for fabrication of hand-held devices with robust sensing capabilities suitable for target specific sensing applications in large scale production. Recent strides in miniaturization and development of cost-effective and easy to use units combined with smartphones and portable devices have seen widespread adoption, making it easier for industries and consumers alike to ensure food safety [84].

5 Conclusion

In summary, this chapter demonstrated a brief introduction on food adulteration and the various risks related to it. This chapter also discussed about the principles of visual colorimetric and plasmonic sensing and how their interactions between specific target analytes and nanostructure systems paved the way to a promising field of study. Diverse sensing techniques involving colorimetric and plasmonic in various food adulterant detection which includes artificial dyes, preservatives, additives etc. were also discussed. The discussion further continued into the advantages and challenges involved in such sensing schemes and comparison with other present sensing methods. Lastly, future of nano-based sensors as a promising technology were explored in food monitoring and safety.

Acknowledgments Author UD would like to acknowledge Department of Science and Technology (DST), India for providing financial assistance in form of DST INSPIRE Fellowship [IF190914]. NKD would like to acknowledge CSIR, India for the financial support received as CSIR-SRF Fellowship [File No.: 09/0796(12416)/2021-EMR-I].

References

1. Y.Z. Feng, D.W. Sun, Application of hyperspectral imaging in food safety inspection and control: a review. Crit. Rev. Food Sci. Nutr. **52**(11), 1039–1058 (2012)
2. K. Goyal, P. Kumar, K. Verma, Food adulteration detection using artificial intelligence: a systematic review. Arch. Comput. Methods Eng. **29**(1), 397–426 (2022)
3. P. Galvin-King, S.A. Haughey, C.T. Elliott, Herb and spice fraud; the drivers, challenges and detection. Food Control **88**, 85–97 (2018)
4. S. Lohumi, S. Lee, H. Lee, B.K. Cho, A review of vibrational spectroscopic techniques for the detection of food authenticity and adulteration. Trends Food Sci. Technol. **46**(1), 85–98 (2015)
5. S.N. Jha, P. Jaiswal, M.K. Grewal, M. Gupta, R. Bhardwaj, Detection of adulterants and contaminants in liquid foods—a review. Crit. Rev. Food Sci. Nutr. **56**(10), 1662–1684 (2016)
6. C.M.E. Gossner, J. Schlundt, P. Ben Embarek, S. Hird, D. Lo-Fo-Wong, J.J.O. Beltran, A. Tritscher, The melamine incident: implications for international food and feed safety. Environ. Health Perspect. **117**(12), 1803–1808 (2009)
7. A. De Brún, L. Shan, Á. Regan, Á. McConnon, P. Wall, Exploring coverage of the 2008 Irish dioxin crisis in the Irish and UK newsprint media. Health Commun. **31**(10), 1235–1241 (2016)
8. J. Premanandh, Horse meat scandal–A wake-up call for regulatory authorities. Food Control **34**(2), 568–569 (2013)
9. T. Agres, The cumin scandal: accidental or fraudulent. Food Qual. Saf. **430** (2015)
10. R. Valand, S. Tanna, G. Lawson, L. Bengtström, A review of Fourier Transform Infrared (FTIR) spectroscopy used in food adulteration and authenticity investigations. Food Addit. Contam. Part A **37**(1), 19–38 (2020)
11. D.I. Ellis, V.L. Brewster, W.B. Dunn, J.W. Allwood, A.P. Golovanov, R. Goodacre, Fingerprinting food: current technologies for the detection of food adulteration and contamination. Chem. Soc. Rev. **41**(17), 5706–5727 (2012)
12. K.M. Mayer, J.H. Hafner, Localized surface Plasmon resonance sensors. Chem. Rev. **111**(6), 3828–3857 (2011)
13. M.I. Stockman, Nanoplasmonics: past, present, and glimpse into future. Opt. Express **19**(22), 22029–22106 (2011)
14. D. Alba-Molina, J.J. Giner-Casares, M. Cano, Bioconjugated plasmonic nanoparticles for enhanced skin penetration. Surf.-Modified Nanobiomater. Electrochem. Biomed. Appl. 219–235 (2020)
15. H.S. Kim, D.Y. Lee, Near-Infrared-Responsive Cancer Photothermal and Photodynamic Therapy Using Gold Nanoparticles. Polymers 10(9), 961–974, (2018)
16. J.F. Li, Z.Y. Li, Manipulation of plasmonic wavefront and light–matter interaction in metallic nanostructures: a brief review. Chin. Phys. B **23**(4), 047305 (2014)
17. M. Mohseni-Dargah, Z. Falahati, B. Dabirmanesh, P. Nasrollahi, K. Khajeh, Machine learning in surface Plasmon resonance for environmental monitoring, in *Artificial Intelligence and Data Science in Environmental Sensing* (Academic Press, 2022), pp. 269–298
18. W. Wang, Y. You, S. Gunasekaran, LSPR-based colorimetric biosensing for food quality and safety. Compr. Rev. Food Sci. Food Saf. **20**(6), 5829–5855 (2021)
19. J.S. Lin, X.D. Tian, G. Li, F.L. Zhang, Y. Wang, J.F. Li, Advanced plasmonic technologies for multi-scale biomedical imaging. Chem. Soc. Rev. **51**(23), 9445–9468 (2022)
20. B. Sepúlveda, P.C. Angelomé, L.M. Lechuga, L.M. Liz-Marzán, LSPR-based nanobiosensors. Nano Today **4**(3), 244–251 (2009)
21. Z. Jin, W. Yim, M. Retout, E. Housel, W. Zhong, J. Zhou, J.V. Jokerst, Colorimetric sensing for translational applications: from colorants to mechanisms. Chem. Soc. Rev. (2024)
22. D. Vilela, M.C. González, A. Escarpa, Sensing colorimetric approaches based on gold and silver nanoparticles aggregation: chemical creativity behind the assay. A review. Analytica chimica acta **751**, 24–43 (2012)
23. C. Salvador-Morales, P. Grodzinski, Nanotechnology tools enabling biological discovery. ACS Nano **16**(4), 5062–5084 (2022)

24. M.T. Alula, L. Karamchand, N.R. Hendricks, J.M. Blackburn, Citrate-capped silver nanoparticles as a probe for sensitive and selective colorimetric and spectrophotometric sensing of creatinine in human urine. Anal. Chim. Acta **1007**, 40–49 (2018)
25. M. Bhattacharya, I.D. Paul, L. Jayachandran, S. Halder, S. Banerjee, Nanobiosensors: principles, techniques, and innovation in nanobiosensors, in *Biosensors in Food Safety and Quality* (CRC Press, 2022), pp. 85–110
26. H. Pan, J. Feng, G.X. He, C.E. Cerniglia, H. Chen, Evaluation of impact of exposure of Sudan azo dyes and their metabolites on human intestinal bacteria. Anaerobe **18**(4), 445–453 (2012)
27. R. Saini, K. Choudhary, Toxic potential of azo dyes: a broader understanding. Hazard. Chem. **469** (2025)
28. L.P. Wu, Y.F. Li, C.Z. Huang, Q. Zhang, Visual detection of Sudan dyes based on the Plasmon resonance light scattering signals of silver nanoparticles. Anal. Chem. **78**(15), 5570–5577 (2006)
29. R. Walker, Nitrates, nitrites and N-nitrosocompounds: a review of the occurrence in food and diet and the toxicological implications. Food Addit. Contam. **7**(6), 717–768 (1990)
30. P. Singh, N. Gandhi, Milk preservatives and adulterants: processing, regulatory and safety issues. Food Rev. Intl. **31**(3), 236–261 (2015)
31. D. Li, Y. Ma, H. Duan, W. Deng, D. Li, Griess reaction-based paper strip for colorimetric/fluorescent/SERS triple sensing of nitrite. Biosens. Bioelectron. **99**, 389–398 (2018)
32. Y. Xiong, M. Li, H. Liu, Z. Xuan, J. Yang, D. Liu, Janus PEGylated gold nanoparticles: a robust colorimetric probe for sensing nitrite ions in complex samples. Nanoscale **9**(5), 1811–1815 (2017)
33. A. Mathaweesansurn, E. Detsri, A new colorimetric method for determination of formaldehyde in sea food based on anti-aggregation of gold nanoparticles. J. Food Compos. Anal. **114**, 104802 (2022)
34. M. Agharkar, S. Mane, Utilization of gold nanoparticles to detect formalin adulteration in milk. Mater. Today: Proc. **45**, 4421–4423 (2021)
35. K.B. Teodoro, F.L. Migliorini, W.A. Christinelli, D.S. Correa, Detection of hydrogen peroxide (H_2O_2) using a colorimetric sensor based on cellulose nanowhiskers and silver nanoparticles. Carbohyd. Polym. **212**, 235–241 (2019)
36. P. Chawla, R. Kaushik, V.S. Swaraj, N. Kumar, Organophosphorus pesticides residues in food and their colorimetric detection. Environ. Nanotechnol. Monit. Manage. **10**, 292–307 (2018)
37. D. Kwon, W. Lee, W. Kim, H. Yoo, H.C. Shin, S. Jeon, Colorimetric detection of penicillin antibiotic residues in pork using hybrid magnetic nanoparticles and penicillin class-selective, antibody-functionalized platinum nanoparticles. Anal. Methods **7**(18), 7639–7645 (2015)
38. J.C. Gukowsky, C. Tan, Z. Han, L. He, Cysteamine-modified gold nanoparticles as a colorimetric sensor for the rapid detection of gentamicin. J. Food Sci. **83**(6), 1631–1638 (2018)
39. U. Das, R. Hoque, R. Biswas, Biosynthesised silver nanoparticles as an efficient colorimetric sensor towards detection of melamine. Appl. Phys. A **129**(5), 328 (2023)
40. U. Das, R. Biswas, N. Mazumder, One-pot interference-based colorimetric detection of melamine in raw milk via green tea-modified silver nanostructures. ACS Omega (2024)
41. S.J. Dutta, G. Chakraborty, V. Chauhan, L. Singh, V.S. Sharanagat, V.K. Gahlawat, Development of a predictive model for determination of urea in milk using silver nanoparticles and UV–Vis spectroscopy. LWT **168**, 113893 (2022)
42. K. Chaisiwamongkhol, S. Labaidae, S. Pon-in, S. Pinsrithong, T. Bunchuay, A. Phonchai, Smartphone-based colorimetric detection using gold nanoparticles of sibutramine in suspected food supplement products. Microchem. J. **158**, 105273 (2020)
43. M.E. Ali, U. Hashim, S. Mustafa, Y.B. Che Man, K.N. Islam, Gold nanoparticle sensor for the visual detection of pork adulteration in meatball formulation. J. Nanomater. **2012**(1), 103607 (2012)
44. D.B. Bittar, T.A. Catelani, L. Pezza, H.R. Pezza, A fast method for the determination of lead in honey samples using stabilizer-free silver nanoparticles. Spectrochim. Acta Part A Mol. Biomol. Spectrosc. **189**, 221–226 (2018)

45. L. Xu, A.M. Abd El-Aty, J.B. Eun, J.H. Shim, J. Zhao, X. Lei, B.D. Hammock, Recent advances in rapid detection techniques for pesticide residue: a review. J. Agric. Food Chem. **70**(41), 13093–13117 (2022)
46. N. Chen, H. Liu, Y. Zhang, Z. Zhou, W. Fan, G. Yu, A. Wu, A colorimetric sensor based on citrate-stabilized AuNPs for rapid pesticide residue detection of terbuthylazine and dimethoate. Sens. Actuators, B Chem. **255**, 3093–3101 (2018)
47. W. Liu, D. Zhang, Y. Tang, Y. Wang, F. Yan, Z. Li, H.S. Zhou, Highly sensitive and selective colorimetric detection of cartap residue in agricultural products. Talanta **101**, 382–387 (2012)
48. S. Wu, D. Li, J. Wang, Y. Zhao, S. Dong, X. Wang, Gold nanoparticles dissolution based colorimetric method for highly sensitive detection of organophosphate pesticides. Sens. Actuators, B Chem. **238**, 427–433 (2017)
49. S. Sharmin, M.A. Yousuf, N. Islam, Multiple fano resonance modes based plasmonic refractive index sensor for edible oil adulteration detection. Optik **312**, 171961 (2024)
50. S. Selvaraj, D. Ravi Shankaran, Nano enabled plasmonic strips for colorimetric detection of food adulterants. Chem. Select **8**(30), e202302027 (2023)
51. I.E. Paul, D.N. Kumar, A. Rajeshwari, S.A. Alex, D. Karthiga, A.M. Raichur, A. Mukherjee, Detection of food contaminants by gold and silver nanoparticles, in *Nanobiosensors* (Academic Press, 2017), pp. 129–165
52. X.Y. Xu, X.G. Tian, L.G. Cai, Z.L. Xu, H.T. Lei, H. Wang, Y.M. Sun, Molecularly imprinted polymer based surface Plasmon resonance sensors for detection of Sudan dyes. Anal. Methods **6**(11), 3751–3757 (2014)
53. N. Borah, D. Gogoi, N.N. Ghosh, C. Tamuly, GA-AuNP@ Tollens' complex as a highly sensitive plasmonic nanosensor for detection of formaldehyde and benzaldehyde in preserved food products. Food Chem. **399**, 133975 (2023)
54. M. Moznuzzaman, M.R. Islam, I. Khan, Effect of layer thickness variation on sensitivity: an SPR based sensor for formalin detection. Sens. Bio-Sensing Res. **32**, 100419 (2021)
55. T. Qi, M. Xu, Y. Yao, W. Chen, M. Xu, S. Tang, H.K. Lee, Gold nanoprism/Tollens' reagent complex as plasmonic sensor in headspace single-drop microextraction for colorimetric detection of formaldehyde in food samples using smartphone readout. Talanta **220**, 121388 (2020)
56. Z. Altintas, Surface Plasmon resonance based sensor for the detection of glycopeptide antibiotics in milk using rationally designed nanoMIPs. Sci. Rep. **8**(1), 11222 (2018)
57. W. Gao, P. Li, S. Qin, Z. Huang, Y. Cao, X. Liu, A highly sensitive tetracycline sensor based on a combination of magnetic molecularly imprinted polymer nanoparticles and surface Plasmon resonance detection. Microchim. Acta **186**, 1–8 (2019)
58. M. Pan, S. Li, J. Wang, W. Sheng, S. Wang, Development and validation of a reproducible and label-free surface Plasmon resonance immunosensor for enrofloxacin detection in animal-derived foods. Sensors **17**(9), 1984 (2017)
59. Y. Xia, R. Su, R. Huang, L. Ding, L. Wang, W. Qi, Z. He, Design of elution strategy for simultaneous detection of chloramphenicol and gentamicin in complex samples using surface Plasmon resonance. Biosens. Bioelectron. **92**, 266–272 (2017)
60. R. Rahad, S.S. Sharar, M.A. Haque, Exploring the diverse applications of plasmonic refractive index sensors: unveiling a new realm of possibilities. Doctoral dissertation, Department of Electrical and Elecrtonics Engineering (EEE), Islamic University of Technology (IUT), Board Bazar, Gazipur-1704, Bangladesh (2023)
61. M.K. Yadav, P. Kumar, R.K. Verma, Detection of adulteration in pure honey utilizing Ag-graphene oxide coated fiber optic SPR probes. Food Chem. **332**, 127346 (2020)
62. D.P. Houhoula, M. Kouzilou, C. Tzogias, V. Kyrana, C. Sflomos, J. Tsaknis, V.P. Lougovois, Effectual gold nanoprobe sensor for screening horse adulteration in meat products (2017)
63. W. Li, Y. Zheng, T. Zhang, S. Wu, J. Zhang, J. Fang, A surface plasmon resonance-based optical fiber probe fabricated with electropolymerized molecular imprinting film for melamine detection. Sensors **18**(3), 828 (2018)
64. S.Y. Oh, M.J. Lee, N.S. Heo, S. Kim, J.S. Oh, Y. Lee, Y.S. Huh, Cuvette-type LSPR sensor for highly sensitive detection of melamine in infant formulas. Sensors **19**(18), 3839 (2019)

65. D. Luna-Moreno, D. Monzón-Hernández, E. Noé-Arias, L.E. Regalado, Determination of quality and adulteration of tequila through the use of surface Plasmon resonance. Appl. Opt. **51**(21), 5161–5167 (2012)
66. A.M. Shrivastav, S.P. Usha, B.D. Gupta, Fiber optic profenofos sensor based on surface Plasmon resonance technique and molecular imprinting. Biosens. Bioelectron. **79**, 150–157 (2016)
67. Y. Saylan, S. Akgönüllü, D. Çimen, A. Derazshamshir, N. Bereli, F. Yılmaz, A. Denizli, Development of surface Plasmon resonance sensors based on molecularly imprinted nanofilms for sensitive and selective detection of pesticides. Sens. Actuators, B Chem. **241**, 446–454 (2017)
68. F. Yılmaz, Y. Saylan, S. Akgönüllü, D. Çimen, A. Derazshamshir, N. Bereli, A. Denizli, Surface Plasmon resonance based nanosensors for detection of triazinic pesticides in agricultural foods, in *New Pesticides and Soil Sensors* (Academic Press, 2017), pp. 679–718
69. F. Yılmaz, N. Bereli, A. Derazshamshir, D. Çimen, S. Akgönüllü, Y. Saylan, A. Denizli, Molecularly imprinted sensors for detecting controlled release of pesticides, in *Controlled Release of Pesticides for Sustainable Agriculture* (2020), pp. 207–235
70. A.R. Sadrolhosseini, S.A. Rashid, H. Soleimani, S. Shafie, A.S.M. Noor, A. Mohammadi, Application of surface Plasmon resonance sensor with polypyrrole chitosan graphene oxide layer to detect the napropamide, in *Journal of Physics: Conference Series*, Vol. 1123, No. 1 (IOP Publishing, 2018), p. 012016
71. S. Miyake, Y. Hirakawa, T. Yamasaki, E. Watanabe, A. Harada, S. Iwasa, H. Narita, Simultaneous detection of six different types of pesticides by an immunosensor based on surface Plasmon resonance. Anal. Sci. **36**(3), 335–340 (2020)
72. B. Liu, J. Zhuang, G. Wei, Recent advances in the design of colorimetric sensors for environmental monitoring. Environ. Sci. Nano **7**(8), 2195–2213 (2020)
73. N. Agrawal, R. Saxena, L. Singh, C. Saha, S. Kumar, Recent advancements in plasmonic optical biosensors: a review. ISSS J. Micro Smart Syst. 1–12 (2022)
74. W. Putzbach, N.J. Ronkainen, Immobilization techniques in the fabrication of nanomaterial-based electrochemical biosensors: a review. Sensors **13**(4), 4811–4840 (2013)
75. K. Narsaiah, S.N. Jha, R. Bhardwaj, R. Sharma, R. Kumar, Optical biosensors for food quality and safety assurance—a review. J. Food Sci. Technol. **49**, 383–406 (2012)
76. S. Othman, N.R. Mavani, M.A. Hussain, N. Abd Rahman, J.M. Ali, Artificial intelligence-based techniques for adulteration and defect detections in food and agricultural industry: a review. J. Agric. Food Res. **12**, 100590 (2023)
77. S. Patari, P. Datta, P.S. Mahapatra, 3d paper-based milk adulteration detection device. Sci. Rep. **12**(1), 13657 (2022)
78. L.A. Putri, Y.D. Prabowo, D.M.M. Dewi, Z. Mumtazah, F.P. Adila, G. Fadillah, H.S. Wasisto, Review of noble metal nanoparticle-based colorimetric sensors for food safety monitoring. ACS Appl. Nano Mater. (2024)
79. T. Li, G. Liu, H. Kong, G. Yang, G. Wei, X. Zhou, Recent advances in photonic crystal-based sensors. Coord. Chem. Rev. **475**, 214909 (2023)
80. D.M.G. Preethichandra, M.D. Gholami, E.L. Izake, A.P. O'Mullane, P. Sonar, Conducting polymer based ammonia and hydrogen sulfide chemical sensors and their suitability for detecting food spoilage. Adv. Mater. Technol. **8**(4), 2200841 (2023)
81. N. Xu, M. Xiao, Z. Yu, B. Jin, M. Yang, C. Yi, On-site quantitation of xanthine in fish and serum using a smartphone-based spectrophotometer integrated with a dual-readout nanosensing assay. Food Chem. **431**, 137107 (2024)
82. S. Lee, S. Kim, D.S. Yoon, J.S. Park, H. Woo, D. Lee, J.H. Lee, Sample-to-answer platform for the clinical evaluation of COVID-19 using a deep learning-assisted smartphone-based assay. Nat. Commun. **14**(1), 2361 (2023)
83. N.H. Anh, M.Q. Doan, N.X. Dinh, T.Q. Huy, D.Q. Tri, B. Van Hao, A.T. Le, Gold nanoparticle-based optical nanosensors for food and health safety monitoring: recent advances and future perspectives. RSC Adv. **12**(18), 10950–10988 (2022)
84. J.L.D. Nelis, A.S. Tsagkaris, M.J. Dillon, J. Hajslova, C.T. Elliott, Smartphone-based optical assays in the food safety field. TrAC, Trends Anal. Chem. **129**, 115934 (2020)

Photonic Crystal Sensor for Fat Measurement in Adulterated Milk

Aditi Lamba, Narendra Kumar, and Kamal Kant Sharma

Abstract This chapter is based on measurement of fat in adulterated milk by optical method, where we choose a binary PC structure with a symmetrically introduced defect. In this structure, the defect layer contains milk of different fat concentrations. We determine the transmission characteristics of such a PC and obtain the defect mode in the PBG region. By changing the fat concentration in milk, it is noted that the defect mode frequency increases linearly showing a prominent role, for an optical sensor for measuring the fat concentration of milk, as the fat concentration of milk sample causes change in refractive index. Such change in refractive index enables shifting in defect mode wavelength along with FWHM and QF. Here, the resonant wavelength of defect mode is considered as indexed parameter for measuring the fat concentration and check the adulteration in milk. Also, the sensitivity or performance of this optical sensor device is high, that is, 233.63 nm/RIU.

Keywords Optical sensor · Milk adulteration · Quality factor · Photonic crystal

1 Introduction

Milk consumption increases continuously as milk is an important part of diet for a large proportion of population worldwide. Hence, with the growing demand, there is a competitive trend in the dairy market in form of complexity of the supply chain owing to high demand. This increased demand of milk is responsible for technological frauds that the producers do in the milk sector and dairy products that become a major problem in many countries. In view of this, a research becomes essential to aware the public about non-intentional carelessness as well as intentional frauds in milk production and so detection techniques are developed and improved time to

A. Lamba · N. Kumar
Department of Physics, SLAS, Mody University of Science and Technology, Lakshmangarh, Sikar, Rajasthan, India

K. K. Sharma (✉)
Department of Botany, Binani Girls College, Bikaner, Rajasthan, India
e-mail: kamalkant128@gmail.com

R. Biswas (ed.), *Optical Techniques for Assessing Food Adulterants*, Progress in Optical Science and Photonics 35, https://doi.org/10.1007/978-981-96-8142-6_11

time. Therefore, we say that milk becomes hazardous to heath due to two causes: contamination and adulteration. In milk contamination, there is unintentional addition of impurities resulting in low quality of milk. They usually occur owing to bad handling practices as well as environmental contamination. On the other-hand, milk adulteration is generally performed intentionally to increase the quantity and hence to degrade the quality of milk for business profit. The most adulteration techniques are made by adding water, detergent, colour and urea to the milk. Milk adulteration causes so many hazards related to health issues.

Other than microbiological quality and safety, the milk quality is standardized on the basis of its nutrient levels mainly protein and fat. Hence, to assess milk quality, the parameters examined are fat, protein, solids, and freezing point as well; while the added adulterants may also improve the value of these parameters in a dishonest manner. The added adulterants are usually starch, wheat flour, urea, cane sugar, and vegetable oil that may preserve its composition. Since milk being a perishable product, so in the summer, it is spoiled during transport. Therefore, some chemical preservatives are added to milk like sodium bicarbonate, hydrogen peroxide, penicillin formaldehyde, and streptopenicillin, etc. The altered consumable liquid milk is diluted to such an extent that, due to its nutritional properties, general public face health problems and malnutrition.

Diluting milk with water increases the volume considerably, which is the most common and simplest method of adulteration by reduction of all milk components, in which chemicals are added to bring the density and color to normal parameters. The dilution of milk with water changes the nutritional, hygienic, and technological quality of the milk. Fat in milk is considered as one of the most valuable components and fraud regarding fat is an important constraint. If quantity of fat is removed, the characteristics of the milk changes as the density increases with increase in dry matter in the milk.

1.1 Methods for Detecting Adulterants in Milk

We know that the adulteration of milk affect and decreases its safety, and the techniques for measuring the adulteration in milk are both qualitative and quantitative in nature. Despite the quantitative detection techniques, the qualitative detection methods involving color change or chemical reactions are more useful because of their simple, fast, and very easy natures [1, 2]. Addition of water in milk decreases the fat per unit volume and so the refractive index of milk changes. Also, several physical and chemical methods have been demonstrated for detecting adulterants in milk [3–5]. Although a wide range of rapid, sensitive, and accurate quantitative methods have been proposed for detection and analysis of milk adulterants, in recent years with the advent of optical methods, photonic crystal based sensors or detectors, can be employed to check the milk concentration, and hence can play important role in fat detection. Therefore, we first focus on the photonic crystal (PC) structures and their various applications and then we design optical sensor to detect the fat concentration.

1.2 Photonic Crystals (PCs)

Photonic crystals are designed periodic nanostructures or media exhibiting photonic bandgaps (PBGs) in frequency ranges of electromagnetic (EM) waves, in an analogy similar to the energy bandgaps in case of electronic motion through a periodic potential, exhibited by periodic pattern of crystals in solid state physics The idea of such electromagnetic media was simultaneously proposed by Yablonovitch and John in the year 1987 [6, 7]. The existence of PBGs is due to multiple scattering at the interfaces and leads to destructive interference of waves. The PC structures are of many types, namely, one–dimensional (1-D), two–dimensional (2-D), and three–dimensional (3-D) that are responsible for existence of PBGs of EM waves propagating through them [8–12]. With the advent of new advance techniques, there exist a large flexibility for choosing the composition of materials with new symmetry of lattice and hence it makes enable to design PCs with new embedded defect structures and consequently they provide an immense potential for tunable optical properties of PCs. Such newly designed PCs offer a potential application in microscopic lasers, optical switches, optical fibers, optoelectronic integrated circuits, and sensors [13–20].

Generally one-dimensional photonic crystals (1-D PCs) easy to fabricate by means of layer by layer coatings and can be exploited in many areas of PC research.

In earlier researches, PCs were proposed to design with dielectrics and semiconductors layers having a good contrast of refractive indices. Despite them, the dispersive media with more tunable characteristics are widely applicable as layers for design of 1DPCs, and novel PC materials are chosen on the basis of their optical characteristics with new prospects. In recent years, many smart materials called novel materials are being used in 1D PCs as a defect or the PC layer and these materials include plasma, superconductor, liquid crystal, chiral, chalcogenide, cold plasma, etc. [21–36]. The PBG in 1-D PCs can be characterized with the variation in several parameters, including effective refractive index contrast, lattice parameter, symmetry, filling fraction, topology, and scalability, in which the material structure is the most effective parameter. Also, photonic crystal system design requires the engineering of the location as well as the bandgap width.

By introducing a designed defect via breaking of the periodicity produces defect mode in PBG [37, 39] with a strong dependence on the defect layer parameters especially thickness and refractive index in which it produces tunable localized modes with the specified defect. The tunable characteristics are externally induced controlled parameters, viz., temperature, pressure, magnetic field, etc. [40–43]. The defect mode tunablity is used in several applications, viz., tunable narrow filter, switching and sensing. Aly et al. [44] proposed the tunable filter application of ultra-violet radiation. Using 1-DPC structure, Zaky et al. [45] discussed the dielectric-superconducting layers for low temperature sensor. Aly et al. [46] exploited the metallic-dielectric PC at ultraviolet region for narrow bandpass filter. Singh et al. [47] proposed a tunable optical sensor with 1-D metallic PC using a liquid crystal defect sandwiched between the layers. Nowadays, 1-D ternary structures are also playing major roles in various optoelectronic applications including sensors. From literature review, we see that optical sensors based on defect structures are widely proposed [48–56].

In this chapter, we proposed a binary PC structure containing a symmetrically introduced defect. In this structure, the defect layer contains milk with fat of different concentrations. We determine the transfer characteristics of the defective PC and determine the defect mode in the zero transmission or PBG region. By changing the concentration of fat in milk viz. the refractive index of the milk, it is noted that the defect mode frequency increases linearly showing a key role for an optical sensor for measuring the fat in an adulterated milk.

2 Theoretical Formulation

2.1 *Proposed 1-D Photonic Crystal Structure with and Without Defect*

In this section, we have introduced a one-dimensional photonic crystal (1-D PC) structure as $(AB)^{10}$ as illustrated in Fig. 1a. Here, it contains two layers of low and high refractive index media such as A and B with 10 times repeatition. For the present study, layer A and B are taken as SiO_2 and TiO_2 materials. The refractive index of SiO_2 and TiO_2 are taken as 1.46 [57] and 2.34 [58] in visible and IR region, respectively. The thickness of SiO_2 and TiO_2 layers are taken as d_1 and d_2, respectively. Layer A plus layer B together are referred to as a unit cell. By repeating this, the entire 1-D PC structure may be created, with the lattice constant being defined as the total thickness $a = d_1 + d_2$. Next, the 1-D PC structured with defect layer is considered in the form, $(AB)^N$ Defect $(BA)^N$, as illustrated in Fig. 1b. The defect layer is sandwiched between two 1-D PC structures, which have mirror symmetry about the defect layer. The layer A and layer B are considered as a 1-D PC structure with same thickness parameters. The number of unit cells on both side of defect layer is taken as N.

2.2 *Theoretical Methodology*

The proposed 1D binary PC with symmetrically introduced defect is depicted above in Fig. 1. We apply the transfer matrix method TMM by multiplying all the layer by layer characteristic matrices. Taking M_1 and M_2 as the characteristic matrices for the layers A and B, respectively; the total characteristics matrix for $(AB)^N$ structure without defect with number of unit cell N, can be considered as [59, 60]

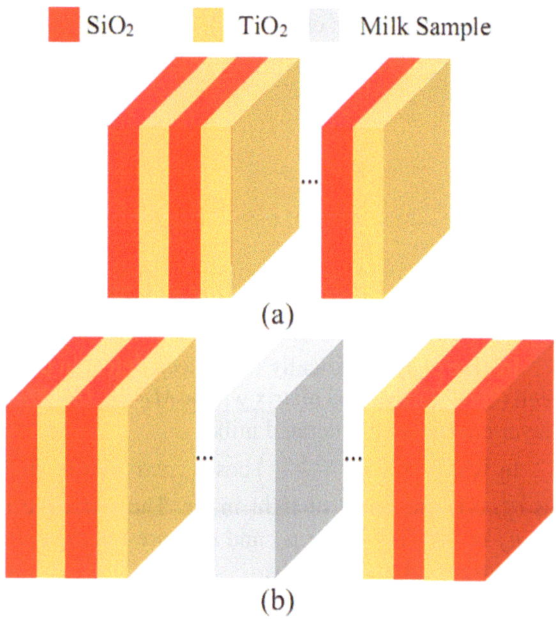

Fig. 1 Schematic diagram of (**a**) 1-D PC structure and (**b**) 1-D PC structure with defect

$$M\,(Nd) = [M\,(a)\,]^N = [M_1 . M_2]^N = \begin{pmatrix} m_{11} & m_{12} \\ m_{21} & m_{22} \end{pmatrix}. \tag{1}$$

Here, the characteristic matrix for each layer is taken as

$$M_p = \begin{pmatrix} \cos\,\gamma_p & -\frac{i}{\varsigma_p}\sin\,\gamma_p \\ -i\,\varsigma_p\,\sin\,\gamma_p & \cos\,\delta_p \end{pmatrix}, \tag{2}$$

The dispersion relation for a periodic structure is expressed as follows [59]:

$$K(\lambda) = \left(\frac{1}{a}\right)\cos^{-1}\left(\frac{m_{11}+m_{22}}{2}\right) \tag{3}$$

where a is the lattice parameter $a = d_1 + d_2$; d_1 and d_2 are the layer thicknesses of the binary PC without defect. By solving the above relation, we get

$$K(\lambda) = \left(\frac{1}{a}\right)\cos^{-1}[\cos(k_1 d_1)\cos(k_2 d_2) + \frac{1}{2}\left(\eta + \frac{1}{\eta}\right)\sin(k_1 d_1)\sin(k_3 d_2)] \tag{4}$$

where $k_1 = \frac{2\pi\,n_1}{\lambda}\cos\theta_1$, and $k_2 = \frac{2\pi\,n_2}{\lambda}\cos\theta_2$.

For the transverse electric (TE) and magnetic (TM) waves, the parameter η is expresses as [60]

$$\eta_{TE} = \frac{k_1}{k_2} \tag{5a}$$

and

$$\eta_{TM} = \left(\frac{n_2}{n_1}\right)^2 \left(\frac{k_1}{k_2}\right) \tag{5b}$$

In case of symmetrically introduced defect in the binary PC, with number of unit cell (2N + 1), the total matrix will be $M_T = (M_A M_B)^N M_D (M_B M_A)^N$; D is the defect layer containing adulterated milk.

In which $\gamma_p = \left(\frac{2\pi . n_p d_p}{\lambda}\right) \cos\theta_p$, and the ray angle in layer-p (p = A, B or D) is θ_p, c is the speed of light in air. The refractive index of p-material is defined as:$n_p = \sqrt{\mu_p \varepsilon_p}$, where μ_p and ε_p represent the permeability and permittivity of a given material, respectively.

Now, we define a parameter ς_p for TE and TM modes as: $\varsigma_p = \sqrt{\frac{\varepsilon_p}{\mu_p}} \cos\theta_p$ and $\varsigma_p = \sqrt{\frac{\mu_p}{\varepsilon_p}} \cos\theta_p$, respectively, with $\cos\theta_p = \sqrt{1 - \frac{n_0^2 \sin^2\theta_0}{n_p^2}}$.

Hence, the transmission coefficient of the 1-DPC is given by [60, 61]

$$t = \frac{2\varsigma_o}{(m_{11} + m_{12}\,\varsigma_s)\,\varsigma_0 + (m_{21} + m_{22}\,\varsigma_s)}; \tag{6}$$

considering $\varsigma_0 = n_0 \cos\theta_0$ and $\varsigma_s = n_s \cos\theta_s$; n_s denotes the substrate refractive index and is assumed to be 1 for the case of air. Finally, the transmittance of the PC can be determined as

$$T = \frac{\varsigma_s}{\varsigma_o} |t|^2. \tag{7}$$

3 Results and Discussion

3.1 Photonic Band Structure

The photonic band structure refers to the relationship between the frequency of light and its wave vector as it propagates through a periodic optical medium, such as a photonic crystal. This concept is analogous to the electronic band structure in solid-state physics, which describes the allowed and forbidden energy levels for electrons

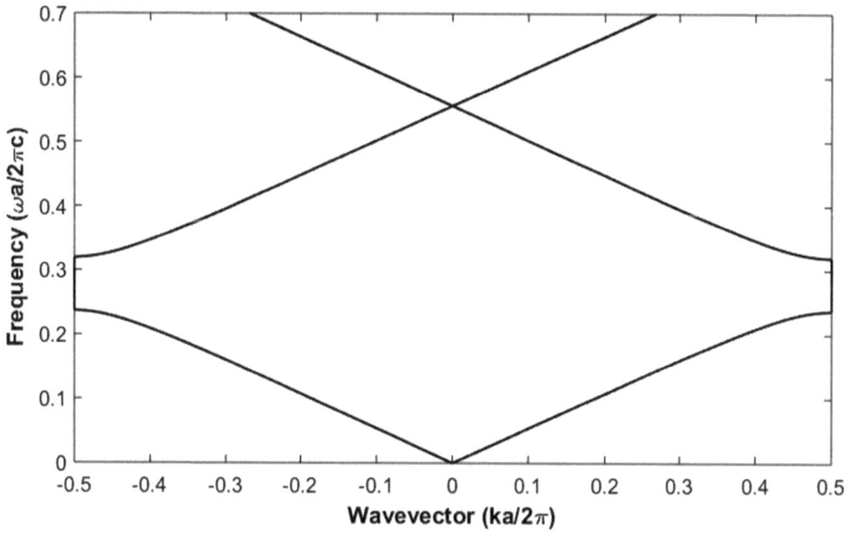

Fig. 2 Photonic band structure

in a crystal lattice. Now, we focused our attention to plot the photonic band structure of 1-D PC structure. In this regard, the thickness of layer A and layer B have to be decided, using the quarter wave condition to maximise the photonic bandgap [10, 62]. So, the thickness of layer A and layer B is considered as $d_1 = \lambda/4n_1$ and $d_2 = d_1 = \lambda/4n_2$, where λ is reference wavelength [62]. Here, we considered the lattice constant ***a*** as unit to plot the photonic band structure of 1-D PC structure on normalised scale as illustrated in Fig. 2.

As it can be seen from Fig. 2 that one photonic band gap is achieved from 0.237 to 0.319 ($\omega a/2\pi c$). Now, the reference wavelength can be considered as 800 nm near IR region for the further study. The thickness of layer A and layer B are achieved as $d_1 = 137$ nm and $d_2 = 85.5$ nm, respectively. Therefore, the lattice constant is obtained as $a = 222.5$ nm. So, the photonic bandgap is obtained from 697.5 nm to 938.8 nm.

3.2 Transmission Spectra

The transmission spectrum of 1-D PC structure $(SiO_2/TiO_2)^{10}$ is plotted as shown in Fig. 3 (upper panel). As can be seen from Figure, a photonic bandgap region is achieved with zero transmission. This photonic bandgap is obtained from 697.5 to 938.8 nm, which are similar with the photonic band structure. Next, we focused our attention to plot the transmission spectrum of 1-D PC structure with defect layer $(SiO_2/TiO_2)^N$/Defect/$(TiO_2/SiO_2)^N$ as shown in Fig. 1b. In this regard, the number of unit cells (N) is taken as 5. The refractive index of defect layer is taken as 1.35

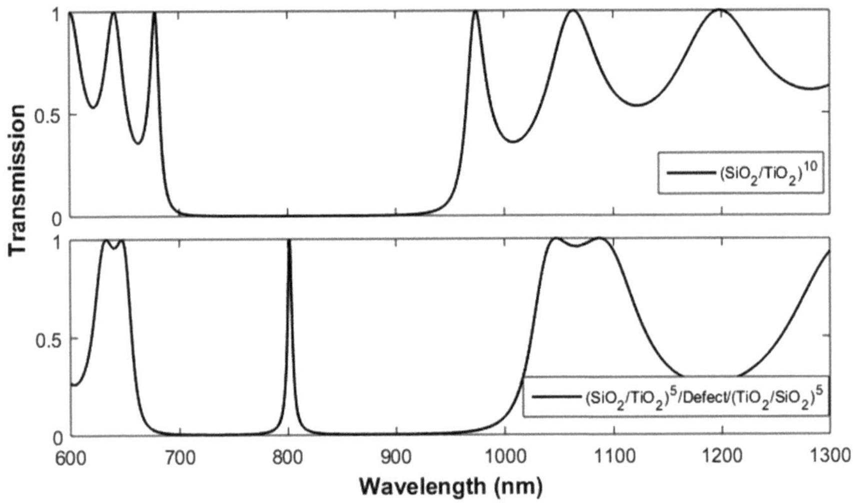

Fig. 3 Transmission spectra for 1D photonic crystal structure

and the thickness of defect layer is taken as 300 nm. The transmission spectrum is plotted as illustrated in Fig. 3 (lower panel).

As can be seen from Fig. 3 that a localised defect mode is achieved in photonic bandgap region and the photonic bandgap region is extended. The defect mode is achieved corresponding to resonance condition of defect layer [10, 62]. The defect mode is obtained at 804 nm, while the bandgap region is obtained from 683.5 to 964.8 nm.

3.3 *Effect of Number of Unit Cells*

Now, we focused our attention to study the effect of number of unit cells (N) on the defect mode. We increased the number of unit cells (N) from 3 to 8 and plotted transmission spectra as illustrated in Fig. 4. The geometrical parameters are taken to be same as previous study and only the number of unit cells (N) is to be varied.

From the Fig. 4, we find that the defect mode gets to be sharper as the number of unit cells increases on both side of defect layer. The position defect mode is to be independent with the number of unit cells. For more detailed study, we calculated the various parameters as transmission (T), resonant wavelength, Full width at half maxima (FWHM) and quality factor (QF) with the number of unit cells and tabulated in Table 1.

As we observe from Table 1 that the transmission is fixed 100% up to the number of unit cells N = 10, thereafter, the transmission decreases sharply with the number of unit cells. This behavior is similar to the increased sharpness of a Fabry–Perot interferometer with increasing reflectivity [63]. The resonant wavelength is shifted

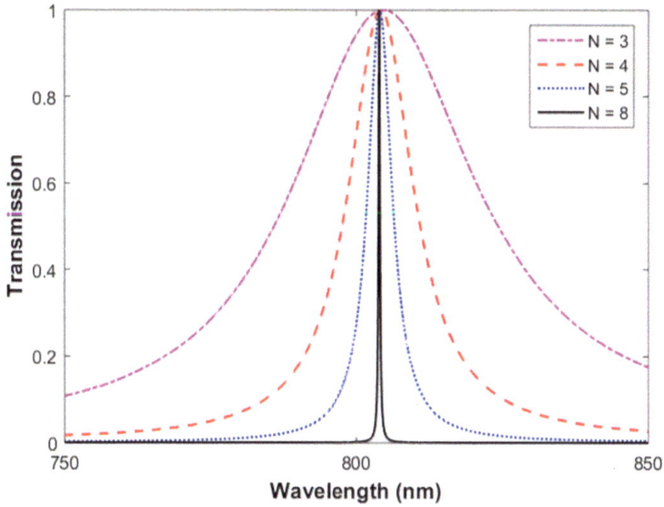

Fig. 4 Transmission spectra with number of unit cells (N)

Table 1 Resonant wavelength, transmission, FWHM and QF with number of unit cells

N	T	Wavelength	FWHM	QF
3	1	804.6309	32.02553	25.12467
4	0.999999	804.1908	12.99869	61.86707
5	0.999986	804.0508	4.978305	161.511
6	0.999927	803.9908	1.917058	419.3878
7	0.999388	803.9508	0.743234	1081.693
8	0.999992	803.9508	0.289139	2780.496
9	0.997173	803.9508	0.113479	7084.575
10	0.962479	803.9508	0.047927	16,774.38
11	0.756283	803.9508	0.02626	30,615.48
12	0.302569	803.9508	0.022439	35,828.46

slightly initially with the number of unit cells up to N = 6 then it is independent with the number of unit cells. The FWHM decreases with the number of unit cells continuously, while QF increases with the number of unit cells [62]. Therefore, the number of unit cells N = 10 can be best value to get high QF values, which can be used for optical sensor application [10].

3.4 Optical Sensor for Milk Adulteration

The defect mode position can be controlled with the geometrical parameters of the defect layer, e.g., thickness or refractive index [10]. If the refractive index of defect layer gets changed, while the thickness of defect layer is kept fixed, then the defect mode will be shifted. This shift in the defect mode results with change in refractive index of defect. Such property of shifting in defect mode is used as optical sensor application of 1-D PC structure with defect layer [62]. Here, the defect layer is considered as milk sample. So, the defect mode can be controlled with the refractive index of milk sample. The transmission spectra of various samples of milk with different fat concentrations are plotted as illustration in Fig. 5. The various parameters of different milk samples are tabulated as illustration in Table 2.

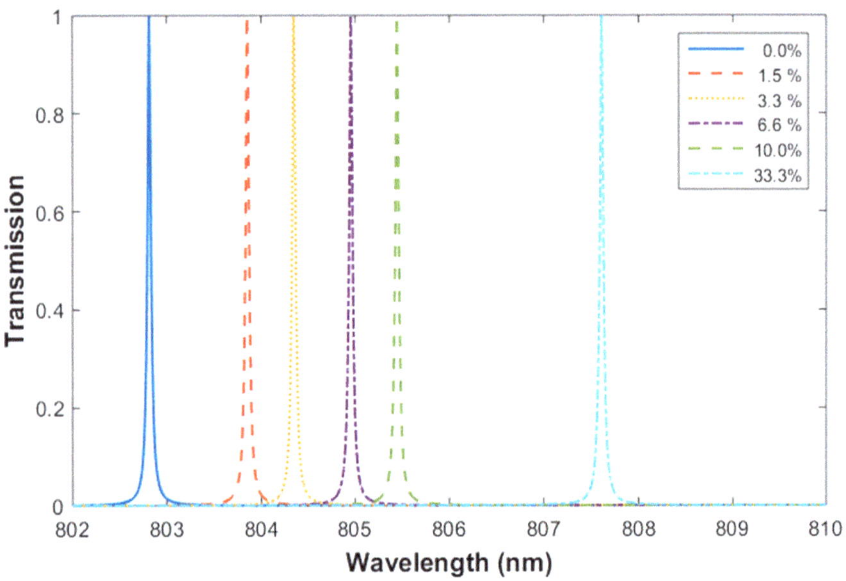

Fig. 5 Transmission spectra for different values of fat concentration

Table 2 Various parameters with fat concentration

Fat concentration (%)	Refractive index	Wavelength (nm)	FWHM	QF
0	1.3452	802.8161	0.04362	18,404.79
1.5	1.3496	803.8523	0.043796	18,354.6
3.3	1.3517	804.3447	0.043917	18,315.09
6.6	1.3543	804.9551	0.044006	18,291.95
10	1.3564	805.445	0.044112	18,258.89
33.3	1.3657	807.6071	0.044582	18,115.14

It is seen from Figure that the defect mode is achieved separately for the different fat concentrations. The shift in the defect mode is a result of change in the refractive index, which is due to fat concentration of milk sample. Similar observations have been reported in previous studies on photonic bandgap structures and their sensitivity to refractive index variations [10, 62, 63]. For the further study, various parameters of transmission peak of defect mode like wavelength, FWHM and QF are calculated with fat concentration of milk and its refractive index values as illustrated in Table 2.

We find from Table 2 that the fat concentration of milk sample causes change in refractive index. Such change in refractive index of defect layer allows shift in defect mode wavelength along with FWHM and QF. Similar optical responses have been observed in photonic crystal-based sensors and biosensing applications [10, 38, 50]. Therefore, the resonant wavelength of defect mode serves as an indexed parameter to measure the fat concentration of milk sample and check the adulteration in milk. The sensitivity is one of the key components to measure the performance of optical sensor device, which can be defined as shift in wavelength per unit refractive index [50, 64]. To evaluate the sensitivity of the proposed optical sensor, we plotted resonant wavelength with the refractive index of defect layer as illustrated in Fig. 6. This approach has been widely used in optical biosensors for detecting small refractive index variations [65].

From Fig. 6, we find that the resonant wavelength of defect mode is shifted linearly with the refractive index of defect layer. We also used curve fitting for the measure data point and find that a straight-line fitting is best suitable. The equation of fitting line is also given in the Fig. 6. The slope of the figure is obtained as 233.63 nm/RIU, which is the sensitivity of the proposed structure.

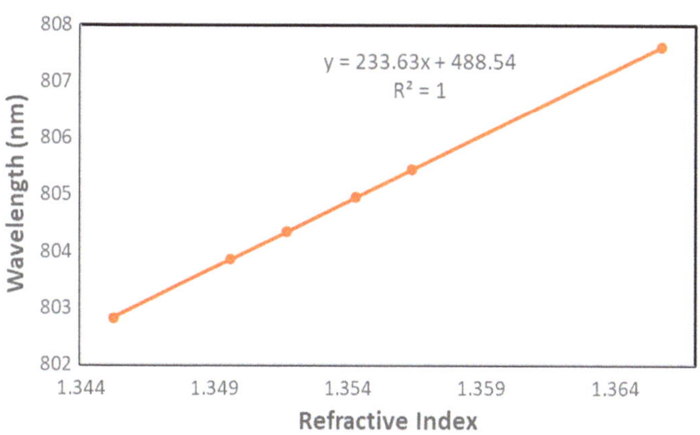

Fig. 6 The resonant wavelength with refractive index

4 Conclusion

The most common and simplest method of adulteration in diluting milk with water reduces all milk components, where chemicals are further added to bring the density and color to normal parameters. It also changes the nutritional, hygienic, and quality of the milk. Fraudulent practices regarding fat tampering is an important constraint as fat is considered as one of the most valuable components. If quantity of fat is removed, the characteristics of the milk changes as the density increases and the refractive index changes. In this chapter, we choose a defect layer containing milk in a 1-D symmetric PC structure, and we obtain the defect mode in its PBG region. It is noted that the defect mode frequency increases linearly showing the features required for an optical sensor for measuring the fat concentration of milk. The resonant wavelength of defect mode can be an indexed parameter for measurement of the fat concentration in adulterated milk. Here, Quality factor and the sensitivity, that is, 233.63 nm/RIU of the designed photonic crystal based optical sensor are high.

References

1. H.M. Kedjia, *Milk Adulteration: Option to Maintain a Quality Product* (Addis Ababa University, Addis Ababa, Ethiopia, 2018)
2. V. Fanelli, I. Mascio, M.M. Miazzi, M.A. Savoia, C. De Giovanni, C. Montemurro, Molecular approaches to agri-food traceability and authentication: an updated review. Foods **10**, 1644 (2021)
3. P. Fisciano, M. Schirone, Food frauds: global incidents and misleading situations. Trends Food Sci. Technol. **114**, 424–442 (2021)
4. C.F. Nascimento, P.M. Santos, E.R. Pereira-Filho, F.R. Rocha, Recent advances on determination of milk adulterants. Food Chem. **221**, 1232–1244 (2017)
5. A.-D. Ionescu, A. Ionut, Cîrîc, M. Begea, A review of milk frauds and adulterations from a technological perspective, a review, MDPI. Appl. Sci. **13**(17), 9821 (2023). https://doi.org/10.3390/app13179821
6. E. Yablonovitch, Phys. Rev. Lett. **58**, 2059 (1987)
7. S. John, Phys. Rev. Lett. **58**, 2486 (1987)
8. P.N. Prasad, *Nanophotonics* (John Wiley & Sons, NJ, 2004)
9. K. Busch, S. John, Phys. Rev. Lett. **83**, 967 (1999)
10. N. Kumar, B. Suthar (eds.), *Advances in Photonic Crystals and Devices* (CRC Press, Taylor & Francis, Boca Raton, USA, 2019). https://doi.org/10.1201/9781351029421
11. M.K. Chhipa, B.T.P. Madhav, B. Suthar, V. Janyani, Ultra-compact with improved data rate optical encoder based on 2D linear photonic crystal ring resonator. Photon Netw. Commun. **44**(1), 30–40 (2022)
12. B. Suthar, A.K. Nagar, A. Bhargava, Tuning the localized mode in point defect chalcogenide photonic crystal. Chalcogenide Lett. **6**(11), 623–627 (2009)
13. N. Kumar, S.P. Ojha, Photonic crystals as infrared broadband reflectors with different angles of incidence: a comparative study. Prog. Electromagnetics Res. PIER **80**, 431–445 (2008). https://doi.org/10.2528/PIER07120502
14. V. Kumar, B. Suthar, A. Kumar, K.S. Singh, A. Bhargava, Design of a wavelength division demultiplexer using Si-based one-dimensional photonic crystal with a defect. Optik **124**(16), 2527–2530 (2013)

15. B. Suthar, A. Bhargava, Tunable multi-channel filtering using 1-D photonic quantum well structures. Prog. Electromagnet. Res. Lett. **27**, 43–51 (2011)
16. V. Kumar, B. Suthar, A. Kumar, K.S. Singh, A. Bhargava, S.P. Ojha, Silicon based one-dimensional photonic crystal as a TM-mode filter. SILICON **6**, 73–78 (2014)
17. S. Ankita, B. Bissa, A. Suthar, Bhargava, High pass filter based on one-dimensional graded photonic crystal. J. Phys: Conf. Ser. **2335**(1), 012006 (2022)
18. M. Radhouene, M. Najjar, M.K. Chhipa, S. Robinson, B. Suthar, Design and analysis a thermo-optic switch based on photonic crystal ring resonator. Optik **172**, 924–929 (2018)
19. A. Bhargava, B. Suthar, Chalcogenide photonic crystals-Exploiting nonlinearity for nanophotonic applications. AIP Conf. Proc. **1536**(1), 15–18 (2013)
20. Z. Gharsallah, M. Najjar, B. Suthar, V. Janyani, Slow light enhanced bio sensing properties of silicon sensors. Opt. Quant. Electron. **51**, 358 (2019)
21. S.P. Ojha, S.K. Srivastava, N. Kumar, S.K. Srivastava, Design of an optical filter using photonic band gap material. Optik **114**(3), 101–105 (2003). https://doi.org/10.1078/0030-4026-00234
22. A. Kumar, N. Kumar, K.B. Thapa, Tunable Broadband reflector and narrowband filter for a dielectric and magnetized cold plasma photonic crystal. The Eur. Phys. J. Plus **133–250**, 1–8 (2018). https://doi.org/10.1140/epjp/i2018-12073-3
23. G.N. Pandey, B. Suthar, N. Kumar, K.B. Thapa, Omnidirectional reflectance of superconductor-dielectric photonic crystal in THz frequency range. J. Supercond. Novel Magn. **34**, 2031 (2021)
24. A. Kumar, N. Kumar, G.N. Pandey, D. Singh, K.B. Thapa, J. Phys.: Condens. Matter **32**, 325701–325713 (2020)
25. B. Suthar, Tuning of guided mode in two dimensional chalcogenide based photonic crystal waveguide. Optik **126**(22), 3429–3431 (2015)
26. B. Suthar, A.K. Nagar, A. Bhargava, Slow light transmission in chalcogenide photonic crystal waveguide. J. Electron. Sci. Technol. **8**(1), 39–42 (2010)
27. N.a Kumar, Sonika, B. Suthar, A. Rostami, Novel optical behaviors of metamaterial and polymer-based ternary photonic crystal with lossless and lossy features. Opt. Commun. **529**(1290), 73 (2023). https://doi.org/10.1016/j.optcom.2022.129073
28. C. Chen, J. Lei, Z. Liu, A ternary seismic metamaterial for low frequency vibration attenuation. Materials **15**, 1246 (2022)
29. A.H. Aly, H.T. Hsu, T.J. Yang, C.J. Wu, C.K. Hwangbo, Extraordinary optical properties of a superconducting periodic multilayer near zero permittivity operation range. J. Appl. Phys. **105**, 083917 (2009). https://doi.org/10.1063/1.3115482
30. B. Suthar, V. Kumar, A. Kumar, K.S. Singh, A. Bhargava, Thermal expansion of photonic band gap for one dimensional photonic crystal. Prog. Electromagnetics Res. Lett. **32**, 81–90 (2012)
31. A. Bhargava, B. Suthar, Localized modes in chalcogenide photonic multilayers with As-S-Se Defect layer. Chalcogenide Lett. **6**(10), 529–533 (2009)
32. N. Kumar, J. Saraf, Tunable reflectance characteristics of magnetized cold plasma based one-dimensional defective photonic crystal. Optik **252**, 168577 (2022)
33. N. Kumar, S. Kaliramna, M. Singh, Design of cold plasma based ternary photonic crystal for microwave applications. SILICON **14**, 6933–6944 (2022)
34. B. Suthar, N. Kumar, S.A. Taya, Design and analysis of tunable multichannel transmission filters with a binary photonic crystal of silver/silicon. The Eur. Phys. J. Plus **137**, 1301 (2022). https://doi.org/10.1140/epjp/s13360-022-03533-6
35. B. Suthar, A. Bhargava, Optical properties of plasma photonic crystals. SILICON **7**(4), 433–435 (2015)
36. A. Kumar, V. Kumar, B. Suthar, K.S. Singh, S.P. Ojha, Nonlinear transmission and reflection characteristics of plasma/polystyrene one dimensional photonic crystal. Optik **125**(1), 393–396 (2014)
37. Ankita, S. Bissa, B. Suthar, A. Bhargava, Optical sensor for biomedical application based on photonic crystal with double defect. Mater. Today: Proc. **62**(8), 5407–5410 (2022)
38. Ankita, S. Bissa, B. Suthar, C. Nayak, A. Bhargava, An improved optical biosensor design using defect/metal multilayer photonic crystal for malaria diagnosis. Res. Opt. **9**, 100304 (2022)

39. Ankita, B. Suthar, S. Bissa, A. Bhargava, Enhancing light manipulation: multidefect photonic crystal filters. J. Opt. **53**, 734–736 (2024)
40. V. Kumar, B. Suthar, A. Kumar, K.S. Singh, A. Bhargava, The effect of temperature and angle of incidence on photonic band gap in a dispersive Si-based one dimensional photonic crystal. Physica B **416**, 106–109 (2013)
41. B. Suthar, A. Bhargava, Pressure sensor based on quantum well-structured photonic crystal. SILICON **13**(6), 1765–1768 (2021)
42. U. Biswas, J.K. Rakshit, J. Das, G.K. Bharti, B. Suthar, A. Amphawan, M. Najjar, Design of an ultra-compact and highly-sensitive temperature sensor using photonic crystal based single micro-ring resonator and cascaded micro-ring resonator. SILICON **13**, 885–892 (2021)
43. A. Kumar, V. Kumar, B. Suthar, A. Bhargava, K.S. Singh, S.P. Ojha, Wide range temperature sensors based on one-dimensional photonic crystal with a single defect. Int. J. Microwave Sci. Technol. **2012**, 182793–182795 (2012)
44. A.H. Aly, S. Barakat, A.F. Amin, Tunable filter based on the 1D photonic crystal within ultraviolet radiations, in *IOP Conference Series Materials Science and Engineering* (2020), p. 012010. https://doi.org/10.1088/1757-899X/956/1/012010.
45. Z.A. Zaky, A.H. Aly, Theoretical study of a tunable low-temperature photonic crystal sensor using dielectric-superconductor nanocomposite layers. J. Supercond. Nov. Magn. **33**, 2983–2990 (2020). https://doi.org/10.1007/s10948-020-05584-1
46. A.H. Aly, H.A. Elsayed, Transmittance properties of one-dimensional metallic-dielectric photonic crystals in near-zero permittivity. Phys. Scr. **94**, 125501 (2019). https://doi.org/10.1088/1402-4896/ab2b7b
47. P. Singh, K.B. Thapa, S.K. Singh, A.K. Gupta, Study of design tunable optical sensor and monochromatic filter of the one-dimensional periodic structure of TiO_2/MgF_2 with defect layer of liquid crystal (LC) sandwiched with two silver layers. Plasmonics **15**, 1845–1854 (2020). https://doi.org/10.1007/s11468-020-01210-x
48. Ankita, B. Suthar, A. Bhargava, Biosensor application of one-dimensional photonic crystal for malaria diagnosis. Plasmonics **16**(1), 59–63 (2021)
49. Ankita, S. Bissa, B. Suthar, A. Bhargava, Graded photonic crystal as improved sensor for nanobiophotonic application. Macromol. Symp. **401**(1), 2100319 (2022)
50. B. Suthar, K. Bhargava, A. Bhargava, Design and analysis of optical biosensor based on symmetric and asymmetric structured one-dimensional photonic crystal for tuberculosis detection. Opt. Eng. **62**(5), 057104–057104 (2023)
51. Z.A. Zaky, H.A. Amer, B. Suthar, A.H. Aly, Gas sensing applications using magnetized cold plasma multilayers. Opt. Quant. Electron. **54**(4), 217 (2022)
52. B. Suthar, A. Bhargava, Enhanced optical sensor for waterborne bacteria-based photonic crystal using graded thickness index. Appl. Nanosci. **13**(8), 5399–5406 (2023)
53. S.A. Taya, I. Colak, B. Suthar, O.M. Ramahi, Cancer cell detector based on a slab waveguide of anisotropic, lossy, and dispersive left-handed material. Appl. Opt. **60**(27), 8360–8367 (2021)
54. N. Kumar, B. Suthar, A. Bhargava, C. Nayak, Analysis of a gas sensor based on one-dimensional photonic crystal structure with a designed defect cavity. Phys. Scr. **98**, 065506 (2023)
55. Ankita, B. Suthar, S. Bissa, A. Bhargava, Revolutionizing optical biosensor with nanocomposite/defect/nanocomposite multilayer 1D photonic crystals. Opt. Quant. Electron. **56**(7), 1116 (2024)
56. Z. Gharsallah, M. Najjar, B. Suthar, V. Janyani, High sensitivity and ultra-compact optical biosensor for detection of UREA concentration. Opt. Quant. Electron. **50**, 249 (2018)
57. L.V. Rodríguez-de Marcos, J.I. Larruquert, J.A. Méndez, J.A. Aznárez, Self-consistent optical constants of SiO_2 and Ta_2O_5 films. Opt. Mater. Express **6**, 3622–3637 (2016)
58. S.V. Zhukovsky, A. Andryieuski, O. Takayama, E. Shkondin, R. Malureanu, F. Jensen, A.V. Lavrinenko, Experimental demonstration of effective medium approximation breakdown in deeply subwavelength all-dielectric multilayers. Phys. Rev. Lett. **115**, 177402 (2015)
59. B. Suthar, A. Bhargava, Enlargement of omni-directional reflection by cascading chalcogenide based photonic crystals. Opt. Commun. **285**(6), 1481–1485 (2012)

60. B. Suthar, A. Bhargava, Temperature-dependent tunable photonic channel filter. IEEE Photonics Technol. Lett. **24**(5), 338–340 (2011)
61. B. Suthar, V. Kumar, K.S. Singh, A. Bhargava, Tuning of photonic band gaps in one dimensional chalcogenide based photonic crystal. Opt. Commun. **285**(6), 1505–1509 (2012)
62. J.D. Joannopoulos, S.G. Johnson, J.N. Winn, R.D. Meade, *Photonic Crystals: Molding the Flow of Light*, 2nd edn. (Princeton University Press, Princeton, NJ, 2008)
63. A. Ghatak, Optics, 6th edn (McGraw-Hill Education, 2017)
64. A.F. Omar, M.Z. Matjafri, Optical sensors for food quality assessment: a review. Sensors **21**(3), 789 (2021)
65. J. Homola, Surface plasmon resonance sensors for detection of chemical and biological species. Chem. Rev. **108**(2), 462–493 (2008)

The Future of Optical Techniques in Food Adulteration Detection

Rajib Biswas

Abstract Food safety and consumer health are seriously threatened by food adulteration worldwide. The future of optical approaches in overcoming this obstacle is examined in this chapter. We explore the possibilities of cutting-edge technology like quantum dots, terahertz spectroscopy, and hyperspectral imaging (HIS) as well as the revolutionary potential of artificial intelligence (AI), including machine learning and deep learning techniques. The importance of multidisciplinary research, teamwork, and tackling issues like data privacy and regulatory compliance are all covered in this chapter. We can create a more robust and sustainable food system and guarantee that everyone has access to wholesome food by utilizing the potential of these cutting-edge technologies.

1 Introduction

Ensuring the safety and authenticity of our food supply is becoming an increasingly significant concern for the global food system. Food adulteration is a serious concern to public health and global food security because it involves the deliberate or inadvertent addition of ingredients to food that lower its quality, mislead customers, or endanger their health. Rapid and effective food safety surveillance is hampered by the time-consuming and labor-intensive laboratory investigations required for traditional methods of adulteration detection. But new developments in artificial intelligence (AI) and optical methods are transforming food safety regulation by providing previously unheard-of chances for quick, precise, and non-destructive adulterant identification [1–3].

The previous chapters had already underlined the importance of different optical techniques. These techniques emerge as versatile with tuning capabilities. The previous chapters highlighted diverse optical techniques for assessing food adulterants. For instance, Chap. 3 prioritized infrared spectroscopy for assessing the same. Likewise, fiber optic-based surface Plasmon resonance biosensors have been dwelt

R. Biswas (✉)
Applied Optics and Photonics, Department of Physics, Tezpur University, Tezpur, India
e-mail: rajib@tezu.ernet.in

R. Biswas (ed.), *Optical Techniques for Assessing Food Adulterants*, Progress in Optical Science and Photonics 35, https://doi.org/10.1007/978-981-96-8142-6_12

upon by Chap. 4. The assessment is primarily based on surface Plasmon resonance as well as localized surface Plasmon resonance. Chapters 3 through 11 delve into the specific applications of various optical techniques in detecting food adulterants.

This chapter reviews the already detailed techniques as discussed in the previous chapters. It explores new technologies and their promise to revolutionize food safety as it dives into the bright future of optical approaches in food adulteration detection. We will look at the state-of-the-art uses of quantum dots, terahertz spectroscopy, and hyperspectral imaging, each of which has special benefits for accurately identifying a variety of adulterants. Additionally, we will examine how AI—including machine learning and deep learning algorithms—can revolutionize the analysis of optical data, improve accuracy, and facilitate real-time food quality monitoring [3–5]. A brief review of the newest technologies, their uses, and the opportunities and problems is highlighted in this chapter.

The importance of multidisciplinary research and cooperation between scientists, engineers, technicians, and food safety authorities will also be covered in this chapter. We can expedite the development and deployment of these cutting-edge technologies by cultivating a collaborative atmosphere, guaranteeing their affordability and accessibility for all parties involved in the global food chain.

Lastly, we will analyze the issues and concerns that need to be resolved in order to properly utilize these developments. These include the need for updated legislative frameworks to keep up with technical improvements, concerns about data privacy and security, and making sure that all stakeholders have fair access to these technologies. We can create a more robust and sustainable food system that guarantees everyone has access to wholesome food by utilizing AI and optical approaches.

2 Emerging Optical Technologies

2.1 Hyperspectral Imaging

Hyperspectral imaging (HSI) has become a potent method for accurately identifying a variety of food product adulterants. HSI gathers hundreds or even thousands of tiny, contiguous spectral bands throughout the electromagnetic spectrum, in contrast to standard imaging, which records data in three color bands (RGB). For every pixel in the picture, this offers a comprehensive dataset of spectral information. Because of their molecular structure and chemical makeup, several substances, such as dietary ingredients and possible adulterants, have distinct spectral signatures, or patterns of light absorption or reflectance. These minute spectral variations can be captured by HSI, enabling the detection and measurement of different adulterants [6–10].

With such exquisite features, HST can facilitate identification of adulterated ingredients. It can be helpful in detecting meat species substitution, water addition, and the presence of harmful additives. Likewise, detection of milk adulteration with water, vegetable oils, or other substances can also be realized. The addition of cheaper concentrates or artificial sweeteners can be captured in fruit juices via HSI. In the same way, presence of adulterants such as cheaper substitutes, dyes, or pesticides can be verified via HSI. Another feather that adheres to HSI is the judging ability towards ripeness, freshness, and the presence of defects—thereby ensuring food quality.

With its great spatial resolution and capacity to record extensive spectrum information, hyperspectral imaging provides a potent and adaptable technique for identifying a variety of food product adulterants. We can create more effective and dependable strategies for guaranteeing food safety and quality by fusing HSI with cutting-edge data analysis approaches.

2.2 Terahertz Spectroscopy

Terahertz (THz) spectroscopy uses electromagnetic waves in the 0.1–10 THz frequency range to examine the structure and composition of materials. For food analysis, this method has various benefits, such as:

Non-destructive Testing: THz radiation can be used for non-destructive analysis without sacrificing sample integrity because it is non-ionizing and typically safe for the majority of food products.

Penetration of Packaging Materials: THz waves may pass through a variety of typical packaging materials, including paper, plastic and thin glass layers, allowing food products to be analyzed without requiring a lot of sample preparation [10–15].

Sensitive to Molecular Vibrations: THz radiation interacts with vibrational modes of molecules to produce distinct spectral fingerprints that can be used to pinpoint and measure certain food product ingredients.

Quick Analysis: THz spectroscopy is appropriate for high-throughput applications in food processing and quality control since it can offer quick and real-time analysis. Towards identification of Food Adulteration, it can cater to adulterants such as edible oil, melamine etc. in dairy products. Again, by examining their distinct spectrum signatures, THz spectroscopy can differentiate between real and fake food items, including honey, olive oil, and spices. Apart from this, THz imaging can be used to find foreign objects embedded in food products, such as glass shards, metal bits, and stones. Summarily, THz spectroscopy, with its ability to penetrate packaging materials and provide rapid, non-destructive analysis, offers a promising avenue for detecting food adulteration and ensuring food safety.

2.3 Quantum Dots

With their distinct optical and electrical characteristics, semiconductor nanocrystals known as quantum dots (QDs) hold great promise as tools for the very sensitive and targeted detection of food contamination. Here's how:

Size-Tunable Fluorescence: QDs have size-dependent fluorescence, which allows for fine tuning of their emission wavelength through size control during synthesis. This makes it possible to produce QDs that emit light at particular wavelengths, allowing for the simultaneous multiplexed detection of several pollutants.

High Quantum Yield: QDs efficiently transform absorbed light into emitted light due to their high quantum yield. Strong fluorescence signals are produced as a result, making it possible to detect even minute levels of pollutants.

High photostability: QDs are resistant to photobleaching, which is the gradual decrease in fluorescence intensity caused by constant illumination. This enhances the sensitivity and reliability of detection by enabling repeated analysis and long-term observations.

Surface Functionalisation: A variety of molecules, including aptamers, antibodies, and certain ligands, can readily functionalize QDs. This improves the specificity of detection by enabling the targeted targeting of particular pollutants, such bacteria, poisons, or pesticides.

To generate extremely sensitive and specific immunoassays for the detection of foodborne pathogens such Salmonella, E. coli, and Listeria, QDs can be conjugated with antibodies. Traces of mycotoxins, insecticides, and other dangerous chemicals in food products can be found using QDs. Other sensing techniques or fluorescence resonance energy transfer (FRET) can accomplish this. Heavy metal pollutants like lead, mercury, and arsenic can be found in food goods using QDs. Other sensing techniques or fluorescence quenching can do this [16–19].

3 Integration of Artificial Intelligence

3.1 Machine Learning Algorithms

The Machine learning algorithms bear enormous potential in augmenting the analysis of optical data, improving accuracy and speed. The analysis of optical data produced by different sensing techniques is greatly improved by machine learning algorithms. These algorithms can:

Automate Feature Extraction: Manual feature extraction from optical spectra or images is frequently necessary for traditional methods, which can be laborious and subjective. Relevant characteristics that are suggestive of adulteration, such as spectral peaks, textures, and patterns, can be automatically detected by machine learning algorithms.

Increase Classification Accuracy: Machine learning models can create reliable classifiers that can reliably differentiate between real and tampered food items by learning from sizable datasets of legitimate and tampered samples. When compared to conventional techniques, this can greatly increase the sensitivity and specificity of detection.

Improve Prediction Capabilities: Using the optical data, machine learning models can forecast the kind and degree of adulteration. By reducing the impact of adulteration on food safety and quality, this predictive capability can be useful for early detection and intervention.

Accelerate Analysis: Real-time or nearly real-time analysis of food samples is made possible by machine learning algorithms' ability to analyze massive amounts of data rapidly and effectively. This can greatly speed up the detection process and increase the effectiveness of inspections for food safety [20–23].

3.2 Deep Learning Models

A subset of machine learning called, deep learning, has shown impressive results in the analysis of complicated data, such as spectra and pictures. Convolutional neural networks (CNNs) and recurrent neural networks (RNNs), the two types of deep learning models, can lead to.

Extract High-Level Features: These models are able to automatically learn intricate, hierarchical features from raw optical data, including textures, shapes, and subtle spectral variations, that may be difficult for human experts or conventional machine learning algorithms to detect.

Manage High-Dimensional Data: Hyperspectral images, which comprise hundreds or even thousands of spectral bands, are an example of optical data with high dimensionality that deep learning models can manage efficiently.

Increase Robustness to Noise and changes: Deep learning models can be trained to withstand noise and data changes, including variations in instrumentation, lighting, and sample preparation.

Facilitate Anomaly Detection: Unusual or abnormal patterns in the data can be detected by deep learning models, which may point to the existence of adulterants or other problems with food safety.

Adoption of Deep learning can enable real time applications in food adulteration detection such as image-based defect and contamination identification. For instance, CNNs can detect mould, foreign objects, and other food product flaws. Using spectral signatures, deep learning algorithms are able to reliably classify various food products and identify their geographic origin.

Meanwhile, adulterants, such as dairy and meat products, can be found in trace levels in complex food matrices using deep learning algorithms. Researchers and business experts can create incredibly precise and effective methods for identifying food adulteration by utilizing deep learning, guaranteeing the security and caliber of our food supply [10, 22, 23].

4 Case Studies and Applications

4.1 Real-World Implementations

Present case studies where advanced optical techniques and AI have been successfully implemented in the food industry.

SIFT uses artificial intelligence (AI) and hyperspectral imaging to precisely sort and examine food goods. Defective or contaminated products can be identified and removed thanks to this technology's ability to detect minute changes in colour, texture, and chemical composition. To ensure consumer safety, SIFT systems, for instance, can be used to detect and eliminate foreign materials from fruits and vegetables, such as stones or metal bits [5].

Numerous businesses are using AI-driven solutions that employ machine learning and computer vision to keep an eye for hygienic and safety infractions on food production lines. By identifying irregularities like inadequate hand washing, food left out, and equipment failures, these systems can help to preserve high standards of food safety and avoid infection.

Crop Monitoring: To track crop health, pinpoint stress zones, and improve fertilization and irrigation, drones outfitted with multispectral sensors and artificial intelligence algorithms are employed. This data-driven strategy can minimize environmental effect while greatly increasing crop yields.

Pest and Disease Detection The use of broad-spectrum pesticides can be minimized by using AI-powered picture analysis to identify early indicators of insect infestations and plant illnesses.

Meat Inspection: To guarantee that only premium items are sent to customers, AI-powered systems can examine photos of meat products to identify flaws like bruising, discoloration, and foreign objects.

Analysis of Dairy Products: By combining AI and near-infrared spectroscopy, it is possible to accurately determine the fat, protein, and lactose content of milk and dairy products.

These case studies show how sophisticated optical techniques and artificial intelligence (AI) may be used in a variety of ways in the food business, from increasing sustainability and efficiency to improving food safety and quality. We may anticipate seeing much more inventive and significant uses in the years to come as these technologies develop further [24].

4.2 Collaborative Efforts

Successful use of AI and optical approaches to improve food safety depends on solid stakeholder collaboration. These collaborations are necessary to quicken research and development. One approach is to provide access to state-of-the-art computing resources, cutting-edge technology, and knowledge of AI and machine learning. The other area is to engage research institutions by making available new analytical techniques. Teaching the upcoming generation of researchers of state of art optical techniques can bolster better scientific outcome. Again, availability of field testing and implementation help, access to real-world data, and regulatory guidance can lead to more collaboration amongst Food Safety Authorities. We can speed up the creation and use of creative solutions for a safer and more sustainable food future by encouraging close partnerships between tech firms, academic institutions, and food safety regulators.

5 Challenges and Considerations

5.1 Data Privacy and Security

The enormous data getting generated from the optical techniques with reference to food inspection and quality create a couple of artifacts which can come at the cost of privacy and security. As such, there must be some effective ways to tackle these. Few needs mention.

Managing Big Datasets: AI and optical methods frequently produce enormous volumes of data. It is essential to make sure that this sensitive data is processed, transmitted, and stored securely. To stop data breaches and illegal access, strong cyber security measures are necessary, such as encryption, access limits, and frequent audits.

Safeguarding Confidential Data: Food businesses frequently hold important confidential data, like recipe formulas and manufacturing procedures. To prevent misuse or illegal access to this sensitive data, data protection measures must be put in place.

Consumer Privacy: Protecting the privacy of consumers is crucial. Establishing trust and upholding moral data practices require explicit data collection and usage regulations as well as gaining customers' informed consent.

5.2 Regulatory Compliance

Likewise, there is a need for updated regulations to keep pace with technological advancements.

Keeping Up with Technological Developments: A flexible regulatory framework is necessary to keep up with the quick developments in AI and optical techniques. The special opportunities and difficulties brought about by these technologies might not be sufficiently addressed by current legislation.

Creating Clear rules: The development, verification, and application of these technologies in the food sector require clear standards and rules. These rules ought to cover topics like the interpretation of results, algorithm transparency, and data quality.

Maintaining Regulatory Consistency: To promote the worldwide adoption of these technologies and prevent trade barriers, it is essential to harmonize regulations across various jurisdictions.

5.3 Cost and Accessibility

Supposing the regulatory compliance is accounted for, another crucial direction is the economic implications and the need for making these technologies accessible to all stakeholders.

Economic Implications: Purchasing sophisticated optical equipment and artificial intelligence software can come with a hefty upfront cost. It is essential to make sure that these technologies are affordable and available to all parties' involved especially small and medium-sized businesses (SMEs) in the food sector.

Creating Inexpensive Solutions: Creating technologies that are more accessible and reasonably priced:

Investigating novel funding strategies: Offering incentives and subsidies to SMEs to encourage their adoption of new technologies.

Addressing the Digital Divide: A truly inclusive food safety system requires that all stakeholders, irrespective of their size or location, have fair access to these technologies.

By proactively addressing these issues and concerns, we can guarantee that the reward of cutting-edge optical methods and artificial intelligence are completely grasped while reducing potential hazards and guaranteeing a sustainable and safe food system for everybody.

6 Future Directions

6.1 Interdisciplinary Research

The intricacies of food safety necessitate a multidisciplinary approach that cuts across conventional boundaries. To spur innovation in this area, more multidisciplinary research is essential. Bringing together specialists in computer science, chemistry, food science, engineering, optics, and spectroscopy allows us to take advantage of a variety of viewpoints and areas of knowledge. Accordingly, the key issues can be as follows.

Create new and improved detection techniques: By combining knowledge of nanotechnology and optics, highly sensitive and targeted sensors for identifying traces of adulterants can be produced.

Enhancing artificial intelligence capabilities: Partnerships between food scientists and computer scientists can hasten the creation of AI algorithms that analyze intricate optical data more quickly and accurately.

Bridging the gap between research and real-world applications: Interdisciplinary research can effectively translate cutting-edge research findings into practical solutions for the food industry, so that these technologies become affordable by common mass.

6.2 Sustainability

By detecting minute alterations in food quality, including microbial growth or impending spoiling, optical sensors enable early intervention and minimize food waste. Likewise, hyperspectral imaging can facilitate more effective sorting and reduce the rejection of perfectly edible produce. It can also precisely determine the quality and ripeness of fruits and vegetables. Another immense potential in the direction of sustainable goal is to maintain the soil conditions for effective management of crop health which will eventually be a deterministic factor for food quality. In this direction, optical sensors can enable tracking of crop health, soil conditions, and water use. This will help farmers minimize their environmental effect by optimizing resource inputs like pesticides and fertilizers. More effective and sustainable livestock management can result from the use of optical tools to monitor the health and welfare of animals.

6.3 Global Impact

Better public health outcomes and lower healthcare expenditures worldwide can stem from increased food safety surveillance. This will further curtail costs in the incidence of foodborne infections. Increased consumer trust in the food supply can be fostered. International trade will be further bolstered by stricter food safety laws and enforcement. Having access to reasonably priced and dependable food safety technologies can enable developing nations to enhance their food quality control systems, lowering food waste and guaranteeing that their populations have access to wholesome food. Food shortages can be avoided and their effects on vulnerable groups can be lessened with early detection of foodborne outbreaks. These technologies can help boost food production while reducing environmental impact by enabling more sustainable and efficient agricultural operations. Further, when the dependence on chemical inputs declines, there will be improved resource management—thereby leading to a more sustainable and resilient food system. The sharing of data and best practices in food safety surveillance can foster international collaboration and accelerate the development and implementation of effective food safety strategies. Global adoption of these technologies can contribute to a more harmonized and interconnected global food system. By harnessing the power of optical techniques and AI, we can create a safer, more sustainable, and more equitable global food system that ensures access to safe and nutritious food for all.

7 Conclusion

Future food safety depends on the ongoing advancement and incorporation of state-of-the-art optical methods. The enormous promise of cutting-edge technologies like terahertz spectroscopy, quantum dots, and hyperspectral imaging—all of which have the potential to completely transform adulterant detection—has been examined in this chapter. Food safety surveillance might become a proactive and effective procedure thanks to the combination of these cutting-edge optical techniques and potent AI algorithms, such as machine and deep learning.

The potential advantages are enormous, even though issues like data protection, regulatory compliance, and accessibility must be carefully considered. From sample analysis to reporting, AI-driven automation can expedite the detection process, improving accuracy and speed. Furthermore, by reducing waste and guaranteeing the security of our food supply, these technologies can make a substantial contribution to sustainable food production.

Implementing these developments successfully calls for a multifaceted strategy. It is essential to do interdisciplinary research that promotes cooperation between scientists, engineers, and stakeholders in the food business. To guarantee the broad acceptance and efficient use of new technologies, sustained investment in R&D as well as the creation of strong regulatory frameworks will be necessary.

It is apparent that these developments will have a global impact on public health and food security in the future. We can create a more robust and sustainable food system that guarantees access to everyone to wholesome food by utilizing AI and optical approaches.

References

1. U. Das, R. Biswas, Unravelling optical properties and morphology of plasmonic gold nanoparticles synthesized via a novel green route. Chem. Pap. **77**(6), 3485–3493 (2023)
2. U. Das, R. Hoque, R. Biswas, Biosynthesised silver nanoparticles as an efficient colorimetric sensor towards detection of melamine. Appl. Phys. A Mater. Sci. Process. **129**(5), 328 (2023)
3. https://www.foodengineeringmag.com/articles/102514-ai-based-inspection-system-brings-for-potato-processors. Assessed on 22 Jan 2025
4. M. Pan, Nanomaterial-based optical detection of food contaminants. Foods (2024). https://doi.org/10.3390/foods13040557
5. D.I.F. Meira, A.I. Barbosa, J.N.P. Borges, R.L. Reis, V.M. Correlo, F. Vaz, Recent advances in nanomaterial-based optical biosensors for food safety applications: ochratoxin-A detection, as case study. Crit. Rev. Food Sci. Nutr. (2023). https://doi.org/10.1080/10408398.2023.2168248
6. B. Park, A new horizon for hyperspectral imaging: macro- and micro-scale measurements for food safety. Nir News **24**(6), 12–14 (2013). https://doi.org/10.1255/NIRN.1388
7. M.S. Kim, A.M. Lefcourt, Y.R. Chen, S.W. Kang, Uses of hyperspectral and multispectral laser induced fluorescence imaging techniques for food safety inspection. Key Eng. Mater. (2004). https://doi.org/10.4028/WWW.SCIENTIFIC.NET/KEM.270-273.1055
8. M. Nikzadfar, M. Rashvand, H. Zhang, A. Shenfield, F. Genovese, G. Altieri, A. Matera, I. Tornese, S. Laveglia, G. Paterna, C. Lovallo, O. Mammadov, B. Aykanat, G.C. Di Renzo, Hyperspectral imaging aiding artificial intelligence: a reliable approach for food qualification and safety. Appl. Sci. **14**(21), 9821 (2024). https://doi.org/10.3390/app14219821
9. M. Aqeel, A. Sohaib, M. Iqbal, H. ur Rehman, F. Rustam, Hyperspectral identification of oil adulteration using machine learning techniques. Curr. Res. Food Sci. (2024). https://doi.org/10.1016/j.crfs.2024.100773
10. N. Gul, K. Muzaffar, S. Zubair, A. Shah, A. Assad, H.A. Makroo, B.N. Dar, Deep learning hyperspectral imaging: a rapid and reliable alternative to conventional techniques in the testing of food quality and safety (2024). https://doi.org/10.15586/qas.v16i1.1392
11. H.J. Shin, J. Lim, K.J. Park, G. Ok, State-of-the-art nondestructive high-speed raster scanning inspection for food safety and quality using terahertz refractive index mapping. Food Chem. Adv. **4**, 100685 (2024). https://doi.org/10.1016/j.focha.2024.100685
12. Y. Lin, Y. Wu, R. Fan, C. Zhan, R. Qing, K. Li, Z. Kang, Identification and quantification of adulteration in collagen powder by terahertz spectroscopy—the effect of spectral characteristics on performance is considered. Spectrochim. Acta Part A Mol. Biomol. Spectrosc. **326**, 125183 (2024). https://doi.org/10.1016/j.saa.2024.125183
13. Y.-Y. Li, C. Shi, H. Wang, Z.-D. Liu, Application of terahertz spectroscopy and imaging techniques in food adulteration. DEStech Trans. Environ. Energy Earth Sci. (2017). https://doi.org/10.12783/DTEEES/SSES/ICFSE2016/10698
14. J. Liu, Terahertz spectroscopy and chemometric tools for rapid identification of adulterated dairy product. Opt. Quant. Electron. **49**(1), 1–8 (2017). https://doi.org/10.1007/S11082-016-0848-8
15. M. Karaliunas, I. Dapsys, A. Urbanowicz, G. Vektaris, A. Vektariene, D. Brazinskiene, S. Asadauskas, I. Kašalynas, G. Valušis, Inspection of oils, caffeine containing foods and consumable plant leaves by time-domain THz spectroscopy, in *International Conference on Infrared, Millimeter, and Terahertz Waves* (2019). https://doi.org/10.1109/IRMMW-THZ.2019.8873934

16. L. Xu, D. Lu, Q. Shi, H. Chen, S. Xie, G. Li, Y.B. She, ZnCdSe-CdTe quantum dots: a "turn-off" fluorescent probe for the detection of multiple adulterants in an herbal honey. Spectrochim. Acta Part A Mol. Biomol. Spectrosc. **221**, 117212 (2019)
17. A.S. Sharma, S. Ali, D. Sabarinathan, M. Murugavelu, H. Li, Q. Chen, Recent progress on graphene quantum dots-based fluorescence sensors for food safety and quality assessment applications. Compr. Rev. Food Sci. Food Saf. **20**(6), 5765–5801 (2021)
18. M. Pan, X. Xie, X. Xie, K. Liu, K. Liu, J. Yang, J. Yang, L. Hong, L. Hong, S. Wang, S. Wang, Fluorescent carbon quantum dots-synthesis, functionalization and sensing application in food analysis. Nanomaterials **10**(5), 930 (2020). https://doi.org/10.3390/NANO10050930
19. S.V. Carneiro, J.J.P. Oliveira, V.S.F. Rodrigues, L.M.U. Dutra Fechine, R.A. Antunes, M. Neto, T.A. de Moura, C.L. Cesar, H.F. de Carvalho, A.R. Paschoal, R.M. Freire, P.B. Almeida Fechine, Doped carbon quantum dots/PVA nanocomposite as a platform to sense nitrite ions in meat. ACS Appl. Mater. Interfaces **14**(38), 43597–43611 (2022). https://doi.org/10.1021/acsami.2c09197
20. S. Manh, P. Duc, T. Xuan, T. Ta, H.L.T. Hong, N.D. Thành, Applications of machine learning coupled with computer vision, electronic nose and untargeted analysis for food quality control. Tạp Chí Kiểm Nghiệm và An Toàn Thực Phẩm **7**(3), 313–328 (2024). https://doi.org/10.47866/2615-9252/vjfc.4360
21. Y. Lin, J. Cheng, J. Ma, C. Zhou, D. Sun, Elevating nanomaterial optical sensor arrays through the integration of advanced machine learning techniques for enhancing visual inspection of food quality and safety. Crit. Rev. Food Sci. Nutr. (2024). https://doi.org/10.1080/10408398.2024.2376113
22. T. Ren, Y. Li, Y. Su, S. Ye, C. Zheng, Machine learning-assisted portable microplasma optical emission spectrometer for food safety monitoring. Anal. Chem. (2024). https://doi.org/10.1021/acs.analchem.3c05332
23. R. Goyal, P. Singha, S.K. Singh, Spectroscopic food adulteration detection using machine learning: current challenges and future prospects (2024). https://doi.org/10.1016/j.tifs.2024.104377
24. *Optical Spectroscopy for Fingerprinting Food: A Photonic Tasting* (Springer eBooks, 2022), pp. 111–120. https://doi.org/10.1007/978-3-031-08136-1_18

MIX
Papier aus verantwortungsvollen Quellen
Paper from responsible sources
FSC® C105338

If you have any concerns about our products,
you can contact us on
ProductSafety@springernature.com

In case Publisher is established outside the EU,
the EU authorized representative is:
Springer Nature Customer Service Center GmbH
Europaplatz 3, 69115 Heidelberg, Germany

Printed by Libri Plureos GmbH
in Hamburg, Germany